Probability
Workbook

Probability Workbook

by Deborah J. Rumsey, PhD

Probability Workbook For Dummies®

Published by: **John Wiley & Sons, Inc.,** 111 River Street, Hoboken, NJ 07030-5774, www.wiley.com

The manufacturer's authorized representative according to the EU General Product Safety Regulation is Wiley-VCH GmbH, Boschstr. 12, 69469 Weinheim, Germany, e-mail: Product_Safety@wiley.com.

For general information on our other products and services, please contact our Customer Care Department within the U.S. at 877-762-2974, outside the U.S. at 317-572-3993, or fax 317-572-4002. For technical support, please visit https://hub.wiley.com/community/support/dummies.

Wiley publishes in a variety of print and electronic formats and by print-on-demand. Some material included with standard print versions of this book may not be included in e-books or in print-on-demand. If this book refers to media that is not included in the version you purchased, you may download this material at http://booksupport.wiley.com. For more information about Wiley products, visit www.wiley.com.

Library of Congress Control Number is available from the publisher.

ISBN 978-1-394-36816-7 (pbk); ISBN 978-1-394-36818-1 (ebk); ISBN 978-1-394-36817-4 (ebk)

Printed and bound by CPI Group (UK) Ltd, Croydon, CR0 4YY

C9781394368167_250226

Contents at a Glance

Table of Contents

Introduction

Are you curious about probability? Are you learning about probability in a class? Or maybe you just like to stay one step ahead of the crowd. If any of these situations describe you, and you like hands-on experience, this workbook is for you.

Probability Workbook For Dummies can help you become more comfortable and confident when working on problems in probability. It includes hundreds of practice problems, with answers fully explained at the end of each chapter. It includes introductory material on each subject, ranging from picturing probability with Venn and tree diagrams to identifying different distributions that model real-life scenarios. Some of the distributions discussed include the Poisson, exponential, continuous uniform, geometric, and hypergeometric. There's also information on the normal and binomial distributions and the central limit theorem, as well as problems about probability in gaming to help you understand what's going on behind the scenes of your favorite lottery and casino games.

As a statistics professor who has taught probability to tens of thousands of students and counting, I've got experience. And I pass along many tips and information that can help you become more comfortable with probability. Doing problems is a huge part of your success, and this workbook offers problems on some of the most common techniques and questions in the area of probability.

About This Book

The first priority of this book is to offer high-quality problems for you to work on — and plenty of them — as you move from topic to topic. My hope is that these problems help cut down on the time you spend spinning your wheels to understand probability. This workbook not only gives you the problems, but it also fully explains the solutions, so even if you know the answer, you might also learn a little bit extra.

Each section starts with important introductory information about each topic, followed by a practice problem that I work out so that you can see how it is done, and ends with problems for you to work out right on the page! There's plenty of space to work in the book itself. You focus not only on solving the problem, but also on setting it up correctly and keeping your eye on the big picture. You also work on interpreting your answers and what misconceptions and mistakes to avoid in the process.

As you move through the workbook, you'll gain confidence and experience, two traits that are extremely helpful and necessary in understanding probability. You may also learn a few surprises along the way about things you thought were true about probability that actually aren't!

This workbook is different from other books on probability in many ways. It includes:

>> **Hundreds of problems to work on at your own pace.** All problems have been written by yours truly, a card-carrying member of the "million probability questions writing and grading club." I know what wrong answers can end up on the page, and I go through them all so that you can avoid them.

>> **Workspace for you to work through the problems directly in each section.** No need to flip elsewhere or bring your own paper to work out problems. You can also take notes in those spaces.

>> **Clear, complete explanations of the problems.** You'll not only see what the answer is, but also know how and why the answer is what it is.

>> **Tips, strategies, and warnings based on my many years of experience.** These tips can help you tackle problems quickly and correctly and get the right answer.

>> **An example problem directly followed by the solution.** You can see a fully worked-out example before you dive into the practice problems.

>> **A focus on step-by-step thinking.** By taking each problem one step at a time, you don't get stuck on a problem midway through or start out the wrong way.

>> **A nonlinear approach.** You can start anywhere in the workbook and move around. Any references to other chapters are for your interest only, and not a requirement.

I also used a few conventions while writing this book that you should be aware of:

>> I use an asterisk (*) to indicate a multiplication sign.

>> I use a capital letter (for example, X, Y, Z) to stand for random variables that represent general values, nothing specific. For example, X = exam scores, or Y = the number of times you roll a die until you get a 6.

>> I use small letters (for example, x, y, z) to stand for exact values of those variables. For example, if you roll a die, the result is x. The small letters usually take on numerical values.

Foolish Assumptions

This workbook is written for anyone who wants to learn more about probability and practice it. Little or no experience with probability is necessary. If you are taking a course in probability, you'll be covering topics such as counting rules, permutations and combinations, and models such as the geometric, negative binomial, and hypergeometric — and maybe a bit on playing poker if you're lucky! It's all here.

If you're taking a probability and statistics class, you'll work on problems in probability in its own right as well as probability as it pertains to statistics, such as p-values and tail probabilities for hypothesis tests. If you're taking a straight statistics course, there's a good chance probability will be included, and you can be ready for it with this workbook!

Perhaps you're interested in probability for everyday use. If so, you'll see plenty of everyday life action in this book involving probability, such as strategies for not having to share the winnings in the lottery.

I wrote this book with you, the student, both in class and in life, in mind. And if you are looking for even more information about probability, I recommend checking out *Probability For Dummies* (also published by Wiley and written by yours truly).

Icons Used in This Book

Throughout this book, icons in the margins highlight valuable information that calls for your attention. Here are the icons you'll encounter and a brief description of each.

EXAMPLE

This icon highlights an example problem with a fully worked solution for use as a reference as you work the practice problems. You can quickly locate the example problems by looking for this icon.

REMEMBER

I use this icon for particular ideas that I hope you'll remember long after you read this workbook.

TIP

This icon points out helpful hints, ideas, or shortcuts that save you time or give you alternative ways to think about a particular concept. I also use this icon to "get down to the nitty-gritty," discussing the types of questions your instructor may ask you and why, revealing what instructors really look for in your answers, and giving you a heads-up on the types of errors that really make them nuts (so you can avoid them at all costs).

WARNING

This icon refers to specific ways you may get tripped up while working on a certain kind of problem and how to avoid them. Commit these items to memory while it still doesn't cost you any points (in other words, before the exam takes place).

Beyond the Book

In addition to the abundance of information and guidance related to probability that I provide in this workbook, you get access to even more help and information online at Dummies.com. Check out this book's online Cheat Sheet. Just go to www.dummies.com and search for "Probability Workbook For Dummies Cheat Sheet."

Where to Go from Here

This book isn't linear; it's modular. It doesn't have to be read from beginning to end. You can jump around from place to place as your needs allow. However, I can make some recommendations on where to start:

>> If you're taking a probability or statistics class based on algebra, I recommend starting with Part 1 to get a basic foundation of some of the notation used in probability and set up problems.

>> If you've taken calculus, you can start with Part 4. However, you do not need calculus for this book. The problems and topics I chose to work on can be done with or without calculus. So don't worry if you haven't had calculus; you can still enjoy the entire workbook.

>> If you're interested in gaming and gambling, head to Chapters 5 and 6. You'll find some ideas regarding strategies for these games and probabilities of winning.

1

The Certainty of Uncertainty: Probability Basics

IN THIS PART . . .

Recognize that probability is part of everyday life.

Review probability terms, symbols, and rules.

Identify when events are independent or mutually exclusive.

Use Venn and tree diagrams to organize and find probabilities.

Calculate conditional probabilities with Bayes' theorem.

Chapter **1**

Seeing Probability in Everyday Life

Probability is part of our everyday lives, from checking the weather and deciding what to wear to looking at the stock market and making predictions to seeing something strange happen and saying, "What are the odds of that?" In this chapter, you explore the ways probability appears in everyday life and learn some important terms you'll encounter throughout the rest of the book.

Understanding What Probability Means

Probability is a word that is used all the time, but it has a specific meaning in the statistics and probability world. First, you start with a random *event*, such as flipping a coin. Then you have the *outcome*, or results, of each flip: heads (H) or tails (T). The *sample space* (S) is the set of all possible outcomes of the random event. Then you have an event that is a certain result — or subset of results — in S, and these results are labeled with capital letters like A, B, C, and so on. And the probability of that event is noted by P(A), which is read "P of A." For example, if you flip a coin three times, let A = you get all heads and B = the number of heads is odd. The probability of some event A is the long-term chance that A occurs over many repetitions of your random event. For example, if there is a 40 percent chance of rain, it means that over many days with the same conditions, it rains 40 percent of the time.

Odds and probabilities are different. A *probability* is a number between zero and one, also known as a *proportion*, like 0.50 or 0.01. It is the total number of ways to do a particular item of interest divided by the total number of ways possible. *Odds* is a ratio, either the odds for a particular outcome or the odds against a particular outcome. The odds for a particular item are the number of ways to get the item of interest divided by the number of ways not to get the item of interest. For example, the probability of rolling a die and getting a 6 is ⅙, but the odds in favor of getting a 6 are 1 in 5, since there are five ways to not get a 6: 1, 2, 3, 4, 5, and one way to get a 6. The odds against an outcome are the number of ways to not get the outcome (5) divided by the number of ways to get it (1). So it's 5-to-1 odds against getting a 6.

Probability can apply to an individual or to a group. Using the roll of a die as an example, the percentage of time you get a 6 when you roll a die is $1 \div 6 = 16.67$ percent if you roll it an infinite number of times. However, the chance of getting a 6 when you roll a die once is 1 out of 6, or ⅙. The probability is the same; the interpretation is different. Also, remember that probability applies to the big picture. For example, if the chance of winning from a scratch-off lottery ticket is 1 out of 10, it doesn't mean buying ten tickets guarantees you a win. It means over the course of an infinite number of tickets, 10 percent of them are winners.

This is another way of saying probability is both a long- and short-term value. If the chance of getting a single head on a single roll of a fair die is ½, then the chance is 1 out of 2. But it also means that if you roll the die an infinite number of times, half of your rolls will turn up heads.

Weather is built on long-term probabilities. If the weather reporter announces that there is a 50 percent chance of rain, that means on 50 percent of the days like this one, it rained. Another way to think of it is that there's a 50–50 chance of rain, but that may not help much either. It's more information than nothing, however.

EXAMPLE

Q. Tell whether the following statement is true or false: "Probability is a short-term idea. If you flip a coin ten times, you will get five heads and five tails."

A. False. Probability is a long-term idea. If you flipped the coin an infinite number of times (long-term), you'd get half heads and half tails, but flipping ten times, you could get any combination of ten heads and tails, and they are all equally likely.

 Tell whether the following statement is true or false: "Probability applies to single individuals only."

2   If you flip a coin three times, what is S?

3   If you flip a coin three times, which outcomes are in event A = {results are all the same}?

4   What are the odds in favor of rolling a 5 or 6 on a fair die?

5   What are the odds against rolling a 1, 2, or 3 on a fair die?

Calculating Probabilities

Different methods exist for finding probabilities of events. One approach is to use the subjective method, which involves using your personal beliefs. For example, you may predict that it will rain based on how it looks outside. Another way is to use a simulation, in which you set up a model that includes probabilities and let it run its course many, many times, and see how many times a certain outcome appears. This method is used by meteorologists to predict events such as where a hurricane will make landfall; it's also used by bracketologists during the NCAA basketball tournament to predict a winner.

You could also use math formulas and simple calculations to figure out probabilities (which you will do throughout this book). Simple calculations work for many problems, especially problems where all the outcomes are equally likely. For example, if you flip a coin once, your possible outcomes are heads or tails. Then, assuming the coin is fair, you have P(Heads)= 1 out of 2, or ½, which is the same for P(Tails). If you roll a die and let A = an even number result, you know $P(A) = P(2, 4, 6) = 3$ out of 6 or $\frac{3}{6} = \frac{1}{2}$.

EXAMPLE

Q. Deciding not to bring your lunch to work because someone will probably go out to eat with you is using what type of probability?

A. Subjective probability; it's based on your beliefs.

---

6 Assume you flip a fair coin two times. What is the probability you get the same result on both flips?

7 Suppose that you roll two fair dice. What's the probability of getting the same result both times?

8 Answer yes or no to the following question: Can every probability be calculated?

Avoiding These Probability Misconceptions

Some ideas about probability seem right but are actually incorrect, and some go against our intuition. First, if you have two outcomes, like heads and tails, their probabilities are both 50 percent, so we say the probability of rolling either option is 50–50. It's 50–50 because the coin is fair, and each outcome is equally likely. But not all two-outcome scenarios play out that way. Just because a sample space S has two outcomes doesn't mean they are both 50–50. For example, consider shooting free throws in basketball. Your chance of making a basket from the free-throw line is probably better than mine, but maybe not as good as a professional basketball player.

Another misconception is that patterns like 1-2-3-4-5-6-7 can't occur randomly (like in the lottery), but of course, they can. Not only that, but they also have the same chance of appearing as any other combination. Yet another misconception is that you can be "on a roll" or "in a slump" when playing casino games. While this may be true when you are a baseball player trying to hit the ball, it's not true that you can be in a slump or on a roll in casino games because trials are independent. That means one play doesn't influence the next.

It's interesting how even picking a number between 1 and 10 at random is not really random unless you use what is called a random number generator (a computer program that comes up with random numbers for you — or for the casinos). If you ask people to pick a number between 1 and 10, fewer people pick 1, 10, and 5 because they are on the ends of the spectrum or directly in the middle. More people pick 3 and 7 as these numbers are in the middle of the lower half and the middle of the upper half. The bottom line is that what you believe is random may not be from a probability standpoint.

The same is true for outcomes of ten coin flips. Some people think you should get something like HTHTHTHTHT if you flip a coin ten times, and you'd never get something like HTTTTTTTTTH. But that's the whole point. These outcomes are equally likely because each flip is fair, and the flips are independent of each other. So, any combination has an equal chance of occurring.

EXAMPLE

Q. Tell whether the following statement is true or false: "HHHHHHHHHH would be harder to get on ten coin flips than HTHTHTHTHT."

A. False. They have the same probability.

---

9 Tell whether the following statement is true or false: "As you come to a stoplight, it's either red, green, or yellow. That means each color has ⅓ chance of occurring when you come to it."

10 Tell whether the following statement is true or false: "People are more likely to choose the number 5 when asked to choose a random number between 1 and 10."

---

11 What does it mean for rolls of a single die to be independent?

12 Tell whether the following statement is true or false: "Just because there are two outcomes does not mean the probability is ½ for each one."

Solutions to Problems in Seeing Probability in Everyday Life

1. False. Probability applies to the entire population you are interested in as well. For example, if the chance of getting heads on a coin flip is $\frac{1}{2}$, that means the next flip has a 50 percent chance of being a head, but it also means that if you flip the coin infinitely, you will get 50 percent heads.

2. If you flip a coin three times, S = {HHH, HHT, HTH, THH, TTH, THT, HTT, TTT}. There are $2 * 2 * 2 = 8$ possible outcomes.

3. A = {results are the same on three flips} = {all heads or all tails} = {HHH, TTT}.

4. The odds are 2 to 4 because there are two ways to get a 5 or 6 and four ways to not get a 4 or 6.

5. The odds are 3 to 3 because there are three ways to get 1, 2, or 3 and three ways to not get 1, 2, or 3.

6. S = {outcomes of two flips} = {HH, HT, TH, TT}. A = {results are the same on two flips} = {HH, TT}. $P(A) = P(HH \text{ or } TT) = \frac{2}{4}$ because all outcomes of coin flips are equally likely.

7. S = {outcomes of two dice}= {11, 12, 13, 14, 15, 16, 21, 22, 23, 24, 25, 26, 31, 32, 33, 34, 35, 36, 41, 42, 43, 44, 45, 46, 51, 52, 53, 54, 55, 56, 61, 62, 63, 64, 65, 66}. There are $6 * 6 = 36$ outcomes. A = {results are the same on both dice} = {11, 22, 33, 44, 55, 66} = $\frac{6}{36}$.

8. No. Some probabilities are subjective, which means they depend on the person thinking about the problem. You might say there is a 10 percent chance of rain, while the meteorologist may say it's 20 percent. And someone else might say it's 30 percent. Certain probabilities are mathematical, like coin flips, dice rolls, and card hands, and some aren't.

9. False. Lights are red, yellow, and green for different amounts of time, depending on the intersection, but at a typical stoplight, the green light is on for the longest duration, allowing traffic to flow. The red light is on for a shorter period to indicate that traffic must stop, and the yellow light flashes briefly before the light turns red to warn drivers that the light is about to change.

REMEMBER

Just because there may be three outcomes in a particular scenario does not mean the probability is $\frac{1}{3}$ for each outcome. Don't assume the probability is evenly split across the number of outcomes as with a fair coin (50–50) or fair six-sided dice ($\frac{1}{6}$ for each number).

10. False. According to research, more people pick 3 and 7 than 5. The point is, all ten numbers are not picked with equal probability, even though it may seem that they should. That's why you can't assume that probability-based phenomena are equally likely if human beings are involved in determining the outcomes. If you used a computerized random-number generator, then the outcomes would be equally likely (like lottery number picks).

11. Independent rolls of a die mean the die must be rolled so that the outcomes don't affect each other.

12. True. The two probabilities can be anything as long as they sum to 1.

Chapter **2**

Teaming up with Probability Terms and Rules

Probability comes with its own language, symbols, and rules, and in this chapter, you review the basics of all of those things. First, you start building up set notation, then you apply the notation to various types of probabilities, such as marginal, joint, and conditional probabilities. Then you apply rules to these probabilities to be able to calculate more complex probabilities. At the end, you find out how you know whether two events — like a playing card being a diamond and red — are independent. (In case you're wondering about the playing card, no, they are not independent!)

Building up Your Set Notation

It all starts with some random phenomena: You pull a card from a 52-card deck, you roll a die, you flip a coin. And you have the set of all possible outcomes denoted by S, the sample space. For example, if you flip a coin twice, your sample space is $S = \{HH, HT, TH, TT\}$. Then you have events such as A, B, or C that are subsets of S. For example, $= \{HH, TT\}$; B might be {the number of heads is even} $= \{HH\}$. The empty set is called the null set, and its notation is $\emptyset$; it contains no outcomes.

REVIEWING THE STANDARD 52-CARD DECK

A standard card deck contains 52 cards. Each card is one of 13 denominations: ace, 2, 3, 4, 5, 6, 7, 8, 9, 10, jack, queen, or king. Sometimes the ace is considered the highest denomination, and sometimes it's considered the lowest denomination; it all depends on what type of game you play with the cards. Each card is also labeled with one of four possible suits. The four suits are diamonds (♦), hearts (♥), clubs (♣), and spades (♠). Diamonds and hearts are red cards (their denomination and suit labels are marked in red), and clubs and spades are black (their denomination and suit labels are marked in black). Thirteen cards make up each suit, which makes 26 of the cards red and 26 cards black. The deck has four cards in each denomination. The cards denoted by J, Q, or K (Jack, Queen, or King, respectively) are called face cards or court cards, because on those cards you see the face (and body) of a jack (or prince), a queen, or a king. One deck contains 12 face cards.

From there, you can work with events in four ways:

» A ∪ B, which is called "A union B" and is made up of all the outcomes in A or B or both.

» A ∩ B, which is called "A intersect B" and is composed of all the outcomes in A and B.

» A′, which is called "A complement" and represents everything in S that is not in A.

» A|B, which is called "A given B" and represents a conditional probability. It represents what's left in A, knowing set B has already occurred. For example, if A = {a card from a 52-card deck is an ace} and B = {card is red} then A = {ace of hearts, ace of diamonds, ace of spades, and ace of clubs}, B = {all 26 red cards}, and A|B = {ace of diamonds and ace of hearts} because you know the card has to be red.

Q. Tell whether the following statement is true or false: "The intersection of A and B is a subset of the union of A and B."

A. True. The union of A and B contains everything in A or B or both, so all outcomes in both A and B are included.

EXAMPLE

 Suppose that S = {1 through 10}; A = {1, 2, 3, 4, 5, 6}, and B = {2, 4, 6, 8, 10}.

a. What is the set that makes up A ∪ B?

b. What is the set that makes up A ∩ B?

c. What is the set that makes up (A ∩ B)′?

Suppose that you flip a coin three times. Let A = {first flip is a head} and B = {second flip is a head}.

a. What are the elements in S, A, and B?

b. What outcomes make up the complement of A union B?

c. What outcomes make up the intersection of A and B?

 If you flip a coin three times, which outcomes are in event A = {results are all the same} and B = {two heads and one tail}?

a. What are the elements in S, A, and B?

b. What is the set $A \cup B$?

c. What is the set $A \cap B$?

Calculating Probabilities

A *probability* is a function that takes a set of outcomes and assigns it a number between 0 and 1. The probability for an event A is written P(A) and stated "P of A." For example, if you flip a coin two times, and A = HH, P(A) would be P(HH), which is ¼. You get this because coin flips are all equally likely when the coin is fair; there are 2 * 2 = 4 outcomes in two coin flips, and HH is one of those four. (The other three are HT, TH, and TT.)

You can also find the probability of the union or intersection of A and B, and the complement of event A. You can also find the probability of A∣B. These probabilities are written as $P(A \cup B), P(A \cap B), P(A')$, and $P(A \mid B)$. Each probability is a value between 0 and 1, and the probability of the null set is, by convention, 0.

EXAMPLE

Q. State which probability is larger or whether they are equal: The probability of getting at least two heads on three coin flips or the probability of getting at most two tails.

A. The probability of getting at most two tails is larger. S = {HHH, HHT, HTH, THH, TTH, THT, HTT, TTT}; A = {two or more heads} = {HHH, HHT, HTH, THH}, and because all outcomes in S are equally likely, P(A) = ⁴⁄₈ = ½. Now B = {at most two tails} = {no more than two tails} = {HHH, HHT, HTH, THH, TTH, THT, HTT }, and P(B) = ⁷⁄₈.

4 Suppose that S = {all cards in a 52-card deck}, A = {cards is an ace}, and B = {cards is red}.

   a. What is P(A ∪ B)?

   b. What is P(A ∩ B)?

   c. What is P(A ∩ B)'?

   d. What is P(A|B)?

5 Suppose that you flip a coin three times. Let A = {first flip is a head} and B = {second flip is a head}.

   a. What's the probability of the complement of A union B?

   b. What's the probability of the intersection of A and B?

   c. What is the conditional probability of B given A?

6 If you flip a coin three times, which outcomes are in event A = {result are all the same} and B = {two heads and one tail}?

   a. What is the probability of the set A'?

   b. What is the probability of the set A ∪ B?

   c. What is the probability of the set A ∩ B?

   d. What is the probability of the set A|B ?

Following Probability Rules

The probabilities of each of the set operations you saw in the previous section come with their own rules: a rule for the complement, a rule for the intersection probability, a rule for the union probability, and a rule for the conditional probability $P(A|B)$.

The rule for the complement is $P(A') = 1 - P(A)$.

The rule for the intersection probability $P(A \cap B)$ is $P(A \cap B) = P(A) P(B)$ if A and B are *independent* (A and B do not affect each other). Otherwise, you have $P(A \cap B) = P(A) P(B|A)$. This is the most general form of the equation and the one you use unless you know that A and B are independent.

The rule of the union probability, $P(A \cup B)$, is called the *addition rule*. It goes like this: $P(A \cup B) = P(A) + P(B) - P(A \cap B)$.

The definition of the conditional probability is $P(A|B) = P(A \cap B)/P(B)$.

One way to help you visualize and understand probabilities is to organize them into a contingency table, as shown in Table 2-1.

Table 2-1   Table of Probabilities

	B	B'	Total
A	0.1	0.2	0.3
A'	0.3	0.4	0.7
Total	0.4	0.6	1.0

For example, $P(A \cap B)$ is the intersection of row A and column B, which is 0.1. $P(A \cup B)$ is the probability of everything in row A or column B or both, which is 0.3 for row A, plus 0.4 for column B, minus 0.1 for the intersection ("and") probability, which gives you 0.6. And $P(A|B) = P(A \cap B)/P(B) = 0.1/0.4 = 0.25$.

EXAMPLE

Q. Suppose that $P(A) = \frac{1}{6}$. What is the probability of A complement?

A. $P(A') = 1 - P(A) = 1 - 1\frac{1}{6} = \frac{5}{6}$.

7 Given the data in Table 2-1, find P(B|A).

8 Given the data in Table 2-1, find P(B∩A′).

9 In general, is P(A|B) = P(B|A)?

10 In general, is P(A∩B) = P(B∩A)?

11 In general, when is P(A) = P(A′)?

12 Suppose that P(A) = 0.5, P(B) = 0.4, and P(A∪B) = 0.8. What is P(A∩B)?

13 Suppose that $P(A|B) = 0.3$ and $P(B) = 0.8$. What is $P(A \cap B)$?

14 Suppose that $P(A|B) = 0.7$ and $P(A) = 0.1$. Can you find $P(B)$?

15 Tell whether the following statement is true or false: "$P(A|B) + P(A'|B) = 1$."

Recognizing Independent Events

Events are independent if they do not affect each other. The definition of independent events is $P(A|B) = P(A)$. Knowing B has occurred doesn't change the probability of A. For example, S = {outcomes of two coin flips}; A = {first coin flip is a head}; and B = {second coin flip is a head}. You can imagine that the results of two coin flips don't affect each other, so they are independent.

But here's the probability work. $P(A|B) = P(A \cap B)/P(B)$; $P(A \cap B) = P(\text{head and head}) = \frac{1}{4}$, and $P(B) = P(\text{HH or TH}) = \frac{2}{4}$. So $P(A|B) = \frac{1}{4}$ divided by $\frac{2}{4}$, which is $\frac{1}{2}$. That's the same as $P(A) = $ first coin flip is a head. So A and B are independent.

However, some events are not independent. For example, if you pull a card out of a 52-card deck and you know it's black, it can't be red, it can't be a heart, and it can't be a diamond. In other words, P(Diamond | Black) = 0 and P(Diamond) = $^{13}/_{52}$, so black and diamond are not independent. These events are called *dependent* events.

Q. Suppose that you flip a coin ten times. Are the results independent?

A. Yes. They are independent because the results do not affect each other.

EXAMPLE

Independence is a good thing for calculating probabilities. If you do something over and over the same way every time — like choosing people randomly, or flipping a coin, or rolling a die — your events are independent; you don't need to worry about conditional probabilities, because P(A | B) = P(A) every time.

TIP

---

16   Given S = {outcome of a die roll}, A = {2, 4, 6} and B = {1, 3, 5}. Are A and B independent?

17   Tell whether the following statement is true or false: "If A and B are dependent, then P(A ∪ B) = 0."

18   Tell whether the following statement is true or false: "If P(A ∩ B) = 0, then A and B are dependent."

19   Tell whether the following statement is true or false: "If events are random, then they are independent."

20 Is it always the case that additional information, like knowing B occurred, is *helpful* to you? For example, if you had the opportunity to pay for additional information, would you? If yes, explain. If not, give an example.

21 Tell whether the following statement is true or false: "Choosing 100 people from a large city using a random number generator will result in data that do not affect each other."

 22 Tell whether the following statement is true or false: "If A and B are independent, then $P(A \cup B)$ becomes $P(A) + P(B) - P(A)P(B)$ by the addition rule."

Discerning Mutually Exclusive Events

Events are mutually exclusive, or *disjoint*, if they cannot occur at the same time. Knowing one of the events happened excludes the other event, and vice versa. Events A and B are mutually exclusive if $P(A \text{ and } B) = 0$.

Q. Are mutually exclusive events independent?

A. No. P(A and B) = P(A)P(B|A). In the second probability, P(B|A) = P(B and A)/P(A). Mutually exclusive events cannot occur at the same time, so P(B|A) = 0/P(A) = 0. If they had been independent, P(B|A) would have been P(B), and P(B) would have had to be zero, which it usually isn't.

This is a major source of confusion for students of probability. But if you remember that mutually exclusive events exclude each other, they must affect each other, and if they affect each other, they cannot be independent.

 23 Tell whether the following statement is true or false: "If A and B are mutually exclusive, then P(A|B) = 0."

 24 Tell whether the following statement is true or false: "If A and B are mutually exclusive, then they are dependent."

 25 Tell whether the following statement is true or false: "If A and B are mutually exclusive, then P(A and B) becomes P(A) + P(B) by the addition rule."

Solutions to Problems in Teaming up with Probability Terms and Rules

1. Suppose that A = {1, 2, 3, 4, 5, 6} and B = { 2, 4, 6, 8, 10}.

   a. $A \cup B$ = {1, 2, 3, 4, 5, 6, 8, 10}

   b. $A \cap B$ = {2, 4, 6}

   c. $(A \cap B)'$ = {1, 3, 5, 7, 8, 9, 10}. The answer is everything in S that is not in the intersection of A and B.

2. Suppose that you flip a coin three times. Let A = {first flip is a head} and B = {second flip is a head}.

   a. S = {HHH, THH, HTH, HHT, HTT, THT, TTH, TTT}; A = {HHH, HTH, HHT, HTT}; B = {HHH, THH, HHT, THT}

   b. $(A \cup B)'$ = {TTH, TTT}

   c. $A \cap B$ = {HHH, HHT}

3. If you flip a coin three times, which outcomes are in event A = {result are all the same} and B = {two heads and one tail}?

   a. S = {HHH, HHT, HTH, THH, TTH, THT, HTT, TTT}; A = {HHH, TTT}; B = {HHT, HTH, THH}

   b. $A \cup B$ = {HHH, TTT, HHT, HTH, THH}

   c. $A \cap B = \varnothing$

4. Suppose that S = {all cards in a 52-card deck}; A = {cards is an ace} and B = {cards is red}.

   a. $P(A \cup B)$ = P(card is red or card is an ace) = P(all four aces, and all cards from 2 to king that are diamonds or hearts) = $^{(4 + 24)}\!/_{52}$ = $^{28}\!/_{52}$ or $^{7}\!/_{13}$.

   b. $P(A \cap B)$ = P (card is red and card is an ace) = P(ace of hearts and ace of diamonds) = $^{2}\!/_{52}$ = $^{1}\!/_{26}$.

   c. $P(A \cap B)'$ = the opposite of the probability from the previous problem. You can take a shortcut and take $1 - P(A \cap B)$ as a shortcut using the complement rule, so the answer is $1 - ^{2}\!/_{52}$ = $^{50}\!/_{52}$ or $^{25}\!/_{26}$.

   d. $P(A \mid B)$ = P(card is an ace | card is red) = $^{2}\!/_{26}$ or $^{1}\!/_{13}$. Note that there are 26 red cards because half the cards are red. And there are four aces, two red, and two black. So the answer is $^{2}\!/_{26}$. Notice that this information didn't help you because there are $^{4}\!/_{52}$ aces in the deck, which reduces to $^{1}\!/_{13}$. So $P(A \mid B) = P(A)$, and A and B are therefore independent.

TIP

If you have to find the complement of A, look and see if you have already found the probability of A itself, or if it's easier to find the probability of A. Then all you have to do is take one minus to get the complement!

(5) S = {HHH, HHT, HTH, THH, TTH, THT, HTT, TTT}. Let A = {HTT, HTH, HHT, HHH} and B = {HHH, HHT, THH, THT}.

    a. The complement of A union B = the complement of {HHH, HTT, HTH, HHT, HHH, THH, THT}, which is {TTH and TTT}. The probability of this set is $\frac{2}{8}$ or $\frac{1}{4}$.

    b. $P(A \cap B) = P(HHH, HHT) = \frac{2}{8} = \frac{1}{4}$

    c. $P(B \mid A) = P(B \cap A) / P(A) = \frac{\frac{1}{4}}{\frac{1}{2}} = \frac{1}{2}$

(6) S = {you flip a coin three times} = {HHH, THH, HTH, HHT, HTT, THT, TTH, TTT}; A = {result are all the same} = {HHH, TTT}; and B = {two heads and one tail} = {HHT, HTH, THH}.

    a. $P(A') = 1 - P(HHH, TTT) = 1 - \frac{2}{8} = \frac{6}{8}$ or $\frac{3}{4}$

    b. $P(A \cup B) = P\{HHH, TTT, HHT, HTH, THH\} = \frac{5}{8}$

    c. $P(A \cap B) = P\{no outcomes\} = P(\varnothing) = 0$

    d. $P(A \mid B) = P(A \cap B)/P(B) = \frac{0}{\frac{3}{8}} = 0$

(7) $P(B \mid A) = P(B \cap A)/P(A) = .\frac{1}{3} = .33$

(8) $P(B \cap A') = 0.3$

(9) No. $P(A \mid B)$ is the probability of the elements in A that occur, given that B has occurred. The formula is $P(A \mid B) = P(A \cap B)/P(B)$; note in the formula for conditional probability, you divide by $P(B)$. And $P(B \mid A)$ is the probability of the elements in B that occur, given that A has occurred. The formula is $P(B \mid A) = P(B \cap A)/P(A)$; note that in this formula for conditional probability, you divide by A. Unless $P(A) = P(B)$, the statement is not true.

(10) Yes. It does not matter what order A and B are in. When you intersect them, you still get the same elements.

(11) $P(A) = P(A')$ when $P(A)$ and $P(A')$ are the same, so it is true when both are 0 or when both are $\frac{1}{2}$.

(12) Suppose that $P(A) = 0.5$, $P(B) = 0.4$, and $P(A \cup B) = 0.8$. You find $P(A \cap B)$ by using the addition rule: $P(A \cup B) = P(A) + P(B) - P(A \cap B)$. Fill in what you have and solve for what you need: $0.8 = 0.5 + 0.4 - P(A \cap B)$. So, $P(A \cap B) = 0.1$.

(13) Suppose that $P(A \mid B) = 0.3$ and $P(B) = 0.8$. $P(A \cap B) = P(A \mid B) P(B) = 0.3*0.8 = 0.24$.

$P(A \cap B) = P(A \mid B) P(B)$ is the multiplication rule, and is the same as the definition of probability cross-multiplied on both sides.

REMEMBER

(14) Suppose that $P(A \mid B) = 0.7$ and $P(A) = 0.1$. You can't find $P(B)$. $P(A \mid B) = P(A \cap B) / P(B)$, and you can't split $P(A \cap B)$ into $P(A)P(B)$ because they are not independent. How do you know they aren't independent? Because $P(A \mid B)$ is not the same as $P(A)$.

$P(A \cap B)$ can be split into $P(A)P(B)$ only if A and B are independent. You can check for independence, or it may be stated in the problem. Don't assume A and B are independent without checking.

WARNING

(15) $P(A \mid B) + P(A' \mid B) = 1$ because $P(A \mid B)$ is the part of B that contains A, and $P(A' \mid B)$ is the part of B that contains A'. Think of B as a pie chart, and A and A' as the slices. The slices have to sum to 1 because they make up the whole pie.

(16) S = {outcomes of a die roll}. A = {2, 4, 6} and B = {1, 3, 5}. A and B are dependent because P(A|B) = 0; there is no overlap between A and B. But P(A) = $\frac{3}{6}$ or $\frac{1}{2}$, which is not equal to 0.

(17) False. If A and B are dependent, then P(A∩B) = 0 only if A and B are mutually exclusive. A and B are dependent if they are mutually exclusive, but they may still be dependent without being mutually exclusive. For example, P(A|B) = $\frac{1}{2}$, but P(A) = $\frac{1}{4}$. In this case, A and B are dependent, but $\frac{1}{2}$ = P(A|B) = P(A∩B) / P(B) and P(A∩B) can't be zero.

(18) True. If P(A∩B) = 0, then A and B are mutually exclusive, which is the strongest form of being dependent that there is. They affect each other so much that they exclude each other.

(19) If events are random, then they are independent. This is true. Events that are random do not affect each other by definition.

TIP

In large studies like large surveys, you want to choose the individuals at random, so you have independent results that don't affect each other, and you don't need to use conditional probabilities.

(20) No. Suppose that you want to find the probability that a card drawn from a 52-card deck is red. Your friend tells you in addition that the card is a 2. Did that help? No, it didn't. Because P(Red card) = $\frac{1}{2}$, and P(2|Red) = $\frac{2}{4}$ = $\frac{1}{2}$. If the information you have and the information you are getting are independent, then the additional information you are getting is not helpful.

(21) True. Choosing 100 people from a large city using a random number generator will result in data that do not affect each other. Because the random number generator generates random numbers, the values and the information from the individuals chosen are independent of each other.

(22) True. If A and B are independent, then P(A∪B) becomes P(A) + P(B) – P(A)P(B) by the addition rule. The addition rule is P(A∪B) = P(A) + P(B) – P(A∩B), and P(A∩B) = P(A) P(B) because A and B are independent.

TIP

Note that independence makes it easier to calculate P(A∪B) because you only need P(A) and P(B) to do it. Independence simplifies a great deal of calculations.

(23) True. If A and B are mutually exclusive, then P(A|B) = P(A∩B)/P(B) = 0/P(B) = 0.

(24) True. If A and B are mutually exclusive, then they are dependent because P(A|B) = P(A∩B)/ P(B) = 0/P(B) = 0 is not equal to P(A).

(25) True. If A and B are mutually exclusive, then P(A and B) becomes P(A) + P(B) by the addition rule. The addition rule is P(A∪B) = P(A) + P(B) – P(A∩B), and P(A∩B) = 0 because A and B are mutually exclusive.

IN THIS CHAPTER

» **Exploring Venn diagrams to organize and picture relationships**

» **Using tree diagrams to display and work with conditional probabilities**

» **Finding marginal probabilities using the law of total probability**

» **Calculating conditional probabilities using Bayes' theorem**

Chapter **3**

Picturing Probabilities: Venn Diagrams, Trees, and Bayes' Theorem

S ome probabilities are apparent from the problem, and others you have to get your hands a little dirty to work out. It's often helpful to map out the information you are given in a problem, and use that map to find and solve the probability rules you need. In this chapter, your maps and probability rules include Venn diagrams, tree diagrams, tables, the law of total probability, and Bayes' theorem, which come from the tree diagrams you've set up.

Before you start putting probabilities into picture form, it's important to remember the difference between $P(A|B)$ and $P(B|A)$ and other probabilities. $P(A|B)$ means the probability of A given B has occurred, so B occurred first. And $P(A|B)$ means the probability of B given A has occurred, so A occurred first. Also, recall that $P(A \cap B) = P(B \cap A)$, so you can switch the intersections and also the union probabilities, such as $P(A \cup B) = P(B \cup A)$, but you can't switch the conditional probabilities.

When your problem involves multiple probabilities, it's helpful to first label all the probabilities you do have, and then from there, use those labels throughout the problem. Often, you are given some probabilities, and you are given a question that involves a probability, and there will be a formula between them that connects what you are given to what you are trying to find.

Also, don't forget the multiplication rule, $P(A \cap B) = P(A)P(B \mid A)$, which states that you multiply events in the order in which they occur. Here, A occurred first, then given A occurred, you have B. You can also see the definition of a conditional probability if you divide both sides of the multiplication rule by $P(A)$. You get $P(B \mid A) = \dfrac{P(B \cap A)}{P(A)}$.

Using Venn Diagrams to Organize Relationships and Find Probabilities

A *Venn diagram* is a set of circles, one for each event, set inside a box that represents the sample space S. You may have seen them a time or two before. In this section, you discover how Venn diagrams can help you visualize probabilities and how they can't.

Start with two events A and B, where A = {1, 2, 3, 4, 5, 6}, and B = {5, 6, 7, 8, 9}, and S = {1, 2, 3, 4, 5, 6, 7, 8, 9, 10}. If you look at A ∩ B, you get {5, 6}. If you look at A ∪ B, you get {1, 2, 3, 4, 5, 6, 7, 8, 9}. The Venn diagram of this situation is shown in Figure 3-1.

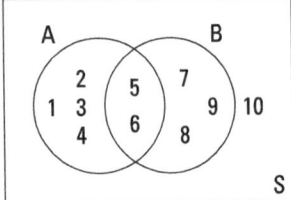

FIGURE 3-1:
A Venn diagram for A and B.

If A and B are independent, then the probability of A intersect B is the product of $P(A)$ times $P(B)$. If they are dependent, then $P(A)$ times $P(B)$ is not $P(A \cap B)$. If A and B are *disjoint* (mutually exclusive, the most extreme sort of dependence), the circles for A and B are separate with no overlap, and $P(A \cap B) = 0$. Finally, A′ (also written as A^c) is everything outside the circle that represents A.

You can also use Venn diagrams to show that certain relationships are true, such as $(A \cup B)^c = A^c \cap B^c$, and $(A \cap B)^c = A^c \cup B^c$. In each case, you make a Venn diagram for the left side of the equation and a Venn diagram for the right side of the equation, and color in the respective parts to show that the colored parts on each side are equal.

However, not all problems can be solved with a Venn diagram. You need the information related to $P(A)$, $P(B)$, and $P(A \cap B)$, or you need the part of A that's in B and not in B, and/or the part of B that's in A and not in A. If you do have those parts, go for the Venn diagram, or you can at least use it to visualize and determine what you still need. Venn diagrams also don't work as well if you are given partial information, such as $P(A \mid B)$, or if there is a sequence to the events, such as A occurs, then $P(B \mid A)$ occurs, and so on.

REMEMBER

Venn diagrams are useful when you're given marginal and intersection (joint) probabilities, and you are asked to find parts, combinations, or complements of those events. They aren't as useful when you have conditional probabilities.

Q. Suppose that a survey shows that 50 percent of college football fans who go to a game will buy food and not merchandise; 20 percent will buy merchandise and not food, and 15 percent will buy both. What percentage will not buy food or merchandise?

EXAMPLE

A. The answer is 0.15. You have $P(F \cap M') = 0.50$, $P(F' \cap M) = 0.20$, and $P(F \cap M) = 0.15$. You want $P(F \cup M)$. This is a perfect problem for Venn diagrams because it gives you intersections, not conditional probabilities. The following figure illustrates what's going on. The phrase "will not buy food or merchandise" in the problem means you want the complement of the union of food and merchandise. In other words, you want the complement of the circles, which is the part outside the circles. And because the total probability of the entire Venn diagram, including the part outside the circles, is 1, the part you want is found by taking $1 - (0.50 + 0.20 + 0.15) = 0.15$.

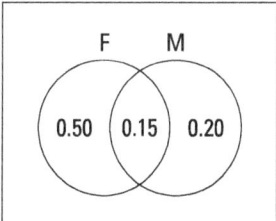

1. Suppose that Petrie is shopping at two stores, Big Store and Big Deal. Petrie buys 50 percent of his items at both stores, 5 percent at Big Store but not Big Deal, and 10 percent at Big Deal but not Big Store. How much does Petrie buy at Big Deal overall?

2. Suppose that Petrie shops for groceries mainly at two stores, Savon and Costless. Petrie buys 50 percent of his items at both stores, 5 percent at Savon but not Costless, and 10 percent at Costless but not Savon. How often does Petrie buy groceries somewhere else besides these two stores?

Suppose that Petrie shops for groceries mainly at two stores, Savon and Costless. Petrie buys 50 percent of his items at both stores, 5 percent at Savon but not Costless, and 10 percent at Costless but not Savon. How much does Petrie buy at Savon or Costless (or both)?

Using Tree Diagrams to Display Marginal and Conditional Probabilities

Tree diagrams are a visual tool that uses lines branching from nodes (or dots) to represent marginal and conditional probabilities. Figure 3-2 illustrates a tree diagram of the marginal probabilities P(A) and P(A′) branching out of the original node. From there, each marginal probability has two nodes branching from it to signify conditional probabilities: P(B|A) and P(B′|A) coming from P(A), and P(B|A′) and P(B′|A′) coming from P(A′).

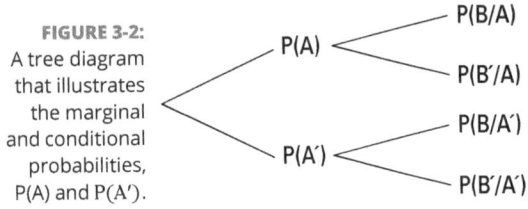

FIGURE 3-2: A tree diagram that illustrates the marginal and conditional probabilities, P(A) and P(A′).

One excellent use of a tree diagram is to calculate intersection probabilities from the information provided in the tree. If you have P(A) and P(B|A) on the first set of branches, you can multiply them using the multiplication rule to get $P(A \cap B)$. You can do this for the other three pathways on the tree. From there, you can find all other intersection probabilities to fill in a 2×2 table with A and A′ in each row and B and B′ in each column.

EXAMPLE

Q. Suppose that the probability that Meriam flies on airline A is 0.60 and the chance she flies on airline B is 0.40. If she flies on airline A, the chance she gets there on time (OT) is 0.90. If she flies on airline B, the chance she gets there on time is 0.80. Use this information to fill in a 2×2 table of intersection (joint) probabilities.

A. The tree diagram of these probabilities is shown in the following figure.

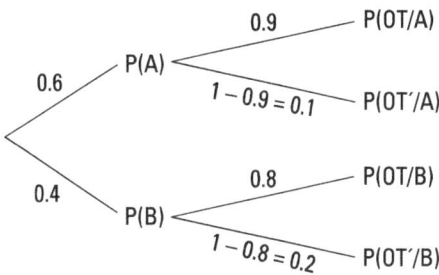

TIP

Note that you can find $P(OT'|A) = 1 - P(OT|A) = 1 - 0.9 = 0.10$ because you know $P(OT|A) + P(OT'|A) = 1$. Same for finding $P(OT'|B) = 1 - 0.8 = 0.2$.

The contents of the four cells of the table are found by multiplying the four sets of pathways that follow the branches of the tree using the multiplication rule. You find the contents of the upper-left cell by finding $P(A \cap OT) = P(A)P(OT|A) = (0.60)(0.90) = 0.54$. The following table shows the answers with the work shown. When all four cells are filled in, you can check your work by confirming that all four probabilities add to 1, which means you've found all the ways to go through the pathways on the tree, as shown in the following table.

	On Time (OT)	Not On Time (OT')
A	0.6 * 0.9 = 0.54	0.6 * 0.1 = 0.06
B	0.4 * 0.8 = 0.32	0.4 * 0.2 = 0.08

 Boris teaches a probability class. Sixty percent of his students in the class are from the statistics department, and 40 percent of his students are from the math department. Of those in the statistics department, 60 percent are women; of those in the math department, 40 percent are women. What percentage of Boris' students are from the statistics department and are women?

Atalie is a photographer. They spend 50 percent of their time taking pictures and 50 percent of their time editing pictures. Of the time Atalie spends taking pictures, they use the computer 10 percent of the time. Of the time Atalie spends editing the pictures, they use the computer 95 percent of the time. What percentage of Atalie's time is spent taking pictures and working on the computer?

6 Restaurant A gets 60 percent of the business in a hotel, and Restaurant B gets the rest of the business in the same hotel. Restaurant A gets an 80 percent satisfaction rate, and Restaurant B gets a 90 percent satisfaction rate. What percentage of all the customers who eat at the hotel eat at Restaurant B and are satisfied?

The Law of Total Probability

The Law of Total Probability helps you figure out a marginal probability when it's not easily apparent from the problem how to get it. Trees play a big role in the Law of Total Probability, as does the multiplication rule. Earlier in Figure 3-2, you saw how P(A) and P(B|A) were laid out on the tree. You also saw three other pathways on the tree that showed P(A) and P(B'|A), P(A') and P(B|A'), and P(A') and P(B'|A'). You saw in the previous section how multiplying these pairs of probabilities gave you the four intersection probabilities needed for a 2×2 table.

You can also use these pairs of probabilities to find the other marginal probabilities in the problem, P(B) and P(B'). You do this through the Law of Total Probability. To use it to find P(B), look back at the tree shown in Figure 3-2 and find the two pathways that end with P(B) and add them together (hence the word "total" in the Law of Total Probability). In this case, out of the four pathways that you find on the tree, if you number them 1, 2, 3, 4 from top to bottom, you add pathway 1 + pathway 3 together to get P(A) P(B|A) + P(A') P(B|A'). Using the multiplication rule twice, this gives you $P(A \cap B) + P(A' \cap B) = P(B)$. (The Venn diagram shown in Figure 3-3 shows how these two parts of B sum up to B.)

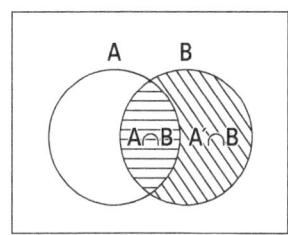

FIGURE 3-3:
How P(A ∩ B) +
P(A' ∩ B) = P(B).

EXAMPLE

Q. Suppose that the probability that Meriam flies on airline A is 0.60 and the chance she flies on airline B is 0.40. If she flies on airline A, the chance she gets there on time is 0.90. If she flies on airline B, the chance she gets there on time is 0.80. What's the total chance she is on time?

A. If you add pathways 1 and 3 from the tree that would go with this problem, you would get $P(A)\,P(OT\,|\,A) + P(B)\,P(OT\,|\,B) = 0.60 * 0.90 + 0.40 * 0.80 = 0.86$. Note that this answer is between the two on-time rates for airlines A and B, which makes sense.

7 Boris teaches a probability class. Sixty percent of his students in the class are from the statistics department, and 40 percent of his students are from the math department. Of those in the statistics department, 60 percent are women; of those in the math department, 40 percent are women. What percentage of all of Boris' students are women?

8 Atalie is a photographer. They spend 50 percent of their time taking pictures and 50 percent of their time editing pictures. Of the time Atalie spends taking pictures, they use the computer 10 percent of the time. Of the time Atalie spends editing the pictures, they use the computer 95 percent of the time. What percentage of Atalie's time is spent working on the computer?

9 Restaurant A gets 60 percent of the business in a hotel, and Restaurant B gets the rest of the business in the same hotel. Restaurant A gets an 80 percent satisfaction rate, and Restaurant B gets a 90 percent satisfaction rate. What percentage of all the customers who eat at the hotel are *not* satisfied?

TIP

Another way to solve Law of Total Probability problems to get marginal probabilities is to take the tree diagram, use the multiplication rule four times on the four pathways to get the four intersection probabilities, then insert them into a 2×2 table. After that, you can just add up the values in the row or column to find the marginal probability you want.

Bayes' Theorem

Bayes' theorem helps you figure out a conditional probability when it's not easily apparent from the problem how to get it. Usually, you are given $P(A|B)$ in the problem and some other information, and you are asked to find $P(B|A)$. Trees play a big role in Bayes' Theorem, as does the multiplication rule and Law of Total Probability.

EXAMPLE

Q. Suppose that the probability that Meriam flies on airline A is 0.60 and the chance she flies on airline B is 0.40. If she flies on airline A, the chance she gets there on time is 0.90. If she flies on airline B, the chance she gets there on time is 0.80. Suppose that she was on time. What's the chance she flew on airline A?

Hint: In this problem, you are given $P(OT|A) = 0.90$, and you are looking for $P(A|OT)$. That's the conditional probability in the opposite order; a clue you need Bayes' theorem.

A. There are four pathways that are found on a tree. If you number them from top to bottom, you have 1, 2, 3, and 4. See the pathways laid out in the following figure.

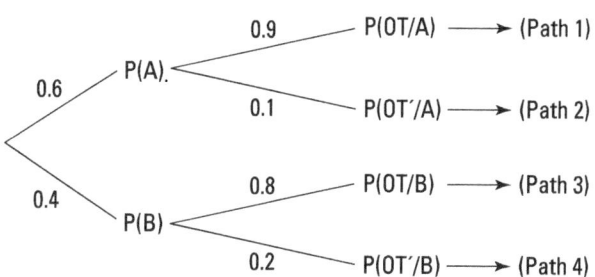

Note that $P(A|OT) = \dfrac{P(A \cap OT)}{P(OT)}$. If you add pathways 1 and 3 from the tree, you get the denominator $P(OT)$: $P(A) P(OT|A) + P(B) P(OT|B) = 0.60 * 0.90 + 0.40 * 0.80 = 0.86$. This is the Law of Total Probability at work. So far, you have the denominator of $P(A|OT)$.

What's more, out of the two pathways in the denominator for getting to be on time (pathway 1 with airline A and pathway 3 with airline B), the one you are interested in is the first one, since it's the probability of A given OT. So, use the multiplication rule to get the numerator of $P(A|OT) = \dfrac{P(A \cap OT)}{P(OT)}$ and you get $P(A)(OT|A)$. The final answer is

$$P(A|OT) = \frac{P(A) P(OT|A)}{P(A) P(OT|A) + P(B) P(OT|B)} = \frac{0.60 * 0.90}{0.60 * 0.90 + 0.40 * 0.80} = \frac{0.54}{0.86} = 0.63.$$ Putting everything together, this is pathway 1 divided by the sum of pathways 1 and 3. This is Bayes' theorem. It has the Law of Total Probability in the denominator and the multiplication rule in the numerator.

TIP

Note that the numerator of Bayes' theorem (pathway 1 in this case) is contained within the denominator of Bayes' theorem (pathways 1 + 3 in this case), so the numerator will always be less than or equal to the denominator, and the final probability will never be greater than 1.

---

10 Boris teaches a probability class. Sixty percent of his students in the class are from the statistics department, and 40 percent of his students are from the math department. Of those in the statistics department, 60 percent are women; of those in the math department, 40 percent are women. Suppose that one of Boris' students is a woman. What's the chance the student came from the math department?

11. Atalie is a photographer. They spend 50 percent of their time taking pictures and 50 percent of their time editing pictures. When taking pictures, Atalie uses the computer 10 percent of the time. When editing the pictures, Atalie uses the computer 95 percent of the time. Suppose that Atalie is working on the computer. What is the chance they are taking pictures?

12. Restaurant A gets 60 percent of the business in a hotel, and Restaurant B gets the rest of the business in the same hotel. Restaurant A gets an 80 percent satisfaction rate, and Restaurant B gets a 90 percent satisfaction rate. Suppose that a customer who eats at the hotel is satisfied. Which restaurant were they more likely to have eaten at, A or B?

Solutions to Problems in Picturing Probabilities: Venn Diagrams, Trees, and Bayes' Theorem

1. The answer is 0.60. BD = Big Deal and BS = Big Store. You want $P(BD) = P(BD \cap BS) + P(BD \cap BS') = 0.50 + 0.10 = 0.60$. The following Venn diagram shows what's going on.

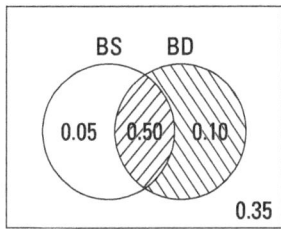

2. The answer is 0.35. S = Savon and C = Costless. You want $P(S \cup C)'$; this is the part outside the two stores. To get this, you can sum everything inside the two circles of the Venn diagram (which represents the union of the two stores), then subtract from 1 to get the complement. You get the following equation and Venn diagram:

$$P(S \cup C)' = 1 - \{P(S \cap C') + P(C \cap S') + P(S \cap C)\} = 1 - (0.05 + 0.10 + 0.50) = 0.35$$

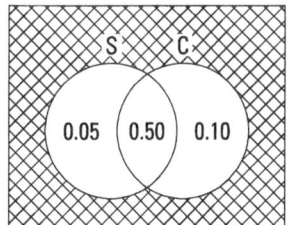

REMEMBER It's important when making Venn diagrams that you remember that all the pieces and parts must sum to one. This includes the part in the box that falls outside both circles, which is easy to forget.

3. The answer is 0.65. This problem is the exact complement of the previous problem, so you take $P(S \cup C)' = 0.35$ and $P(S \cup C) = 1 - 0.35 = 0.65$.

TIP Keep track of previous problems you have done; some later problem might wind up being a complement, and instead of starting from scratch, you can just recognize that and take one minus a previous answer.

4. The answer is 0.36. You are given S = Statistics, M = Math, and W = Woman, and the probabilities are $P(S) = 0.60$; $P(M) = 0.40$; $P(W|S) = 0.60$; and $P(W|M) = 0.40$. You want $P(S \cap W)$. By the multiplication rule, $P(S \cap W) = P(S) P(W|S) = 0.60 * 0.60 = 0.36$.

TIP Note that if the question doesn't involve more than one pathway of the tree in terms of what you want to find, you don't need to create the tree (unless you want to).

5 The answer is 0.05. You know P(Taking) = 0.50 and P(Editing) = 0.50. You are given that P(Computer|Taking) = 0.10 and P(Computer|Editing) = 0.95. You want P(Taking ∩ Computer) = P(Taking) * P(Computer|Taking) = 0.50 * 0.10 = 0.05 by the multiplication rule.

6 The answer is 0.36. You have P(A) = 0.60 and P(B) = 0.40; P(Satisfied|A) = 0.80 and P(Satisfied|B) = 0.90. You want P(B ∩ Satisfied) = P(B) * P(satisfied|B) = 0.40 * 0.90 = 0.36 by the multiplication rule.

7 The answer is 0.52. You have P(Statistics) = 0.60 and P(Math) = 0.40. You also have P(Women|Statistics) = 0.60 and P(Women|Math) = 0.40. You want P(W). The tree diagram for this situation is shown in the following figure.

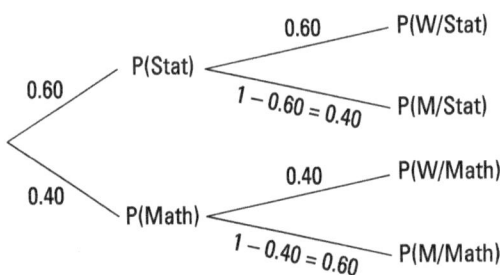

You can see the four pathways on the tree; you need pathway 1 + pathway 3. This is the Law of Total Probability.

P(W) = P(W|Statistics)P(Statistics) + P(W|Math)P(Math) = 0.60 * 0.60 + 0.40 * 0.40 = 0.52.

8 The answer is 0.53. You have P(Taking) = 0.50 and P(Editing) = 0.50. P(Computer|Taking) = 0.10 and P(Computer|Editing) = 0.95. You want P(Computer) overall. The tree diagram is shown in the following figure. You need pathway 1 + pathway 3. This is the Law of Total Probability.

P(Computer) = P(Taking) * P(Computer|Taking) + P(Editing)P(Computer|Editing) = 0.50 * 0.10 + 0.50 * 0.95 = 0.53.

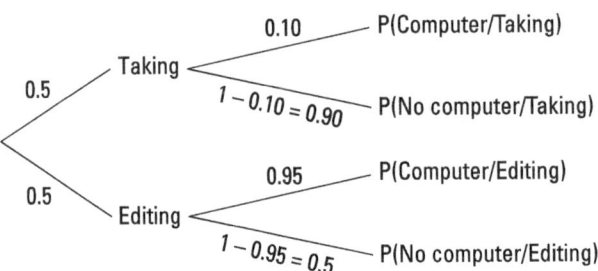

9 The answer is 0.16 or 16 percent. You have P(A) = 0.60 and P(B) = 0.40. P(Satisfied|A) = 0.80 and P(Satisfied|B) = 0.90. You want P(Not Satisfied). Note that you could find P(Satisfied) and take one minus since it's the complement. Or you could fill in the tree and use pathways 2 + 4 to finish the problem. That's what is shown in the tree diagram in the following figure. You have P(Not Satisfied) = P(A)P(not Satisfied | A) + P(B)P(not Satisfied | B) = 0.60 * (1 − 0.80) + 0.40(1 − 0.90) = 0.60 * 0.20 + 0.40 * 0.10 = 0.16 or 16 percent.

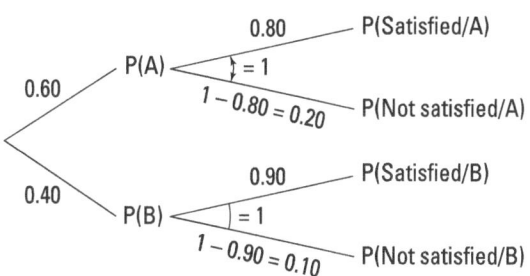

10 The answer is 0.31. You know P(Statistics) = 0.60 and P(Math) = 0.40. P(Women | Statistics) = 0.60 and P(Women | Math) = 0.40. You want P(Math | Women).

What you want is P(Math | Women), and you are given P(Women | Math); that's a clue that you need Bayes' theorem to solve it.

TIP The tree diagram is shown in the following figure. What you want is pathway 3/(pathway 1 + pathway 3). You have P(Math | Woman) = P(Math)P(Woman | Math)/{P(Math)P(Woman | Math) + P(Statistics)P(Woman | Statistics)} = 0.40(0.40)/{(0.40)(0.40)+(0.60)(0.60)} = 0.16/0.52 = 0.31.

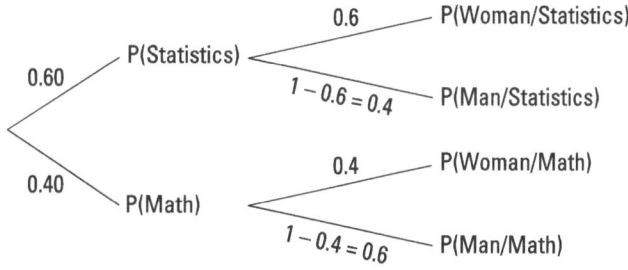

11 The answer is 0.09. You have P(taking) = 0.50 and P(Editing) = 0.50. You also have P(Computer | Taking) = 0.10 and P(Computer | Editing) = 0.95. You want P(Taking | Computer). If you made a tree diagram, you would want pathway 1/(pathway 1 + pathway 3). You have P(Taking | Computer) = P(Taking)(P(Computer | Taking)/{P(Taking)P(Computer | Taking) + P(Editing)P(Computer | Editing)} = 0.50(0.10)/{(0.50)(0.10) + (0.50)(0.95)} = 0.05/0.525 = 0.09.

12 The answer is Restaurant A. P(A) = 0.60 and P(B) = 0.40; P(Satisfied | A) = 0.80 and P(Satisfied | B) = 0.90. You know the person is satisfied, and you want to know which is greater, P(A | Satisfied) or P(B | Satisfied)?

You don't have to figure out both of these since they sum to one. Find one, and the other is one minus that. That's how you can find out which is the largest.

TIP P(A | Satisfied) = P(A)P(Satisfied | A)/{P(A)P(Satisfied | A) + P(B)P(Satisfied | B)} = 0.60(0.80)/{(0.60)(0.80) + (0.40)(0.90)} = 0.48/0.84 = 0.57. That means P(B | Satisfied) = 1 − 0.57 = 0.43. That means they were more likely to have eaten at Restaurant A.

2

Counting on Probability and Betting to Win

Work with contingency tables to explore dependencies and calculate conditional probabilities.

Review how to count the number of permutations and the number of combinations for various scenarios.

Explore the basics of popular casino games of chance and how to avoid Gambler's Ruin.

Chapter **4**

Setting the Contingency Table with Probabilities

A *contingency table* is a table with rows and columns: rows for one event or variable, and columns for another event or variable. You classify the outcomes of the combinations of the events by putting their probabilities in the appropriate row/column combination. In Chapter 3, you saw how to take the values of a tree, multiply them together, and put them into a table format. That was a contingency table.

In this chapter, you work with 2×2 contingency tables, which have two rows and two columns. The rows represent A and A′; the columns represent B and B′. The $2 \times 2 = 4$ intersections in the table are called the *cells* of the table, and the probabilities for the cells are called *joint probabilities*, *intersection probabilities*, or *and probabilities*. You also find out about marginal and conditional probabilities and how to find them using the table. And you use those probabilities to check to see if two events are independent or dependent.

Organizing a Contingency Table

The following is an example of a 2×2 contingency table with events A and B (and their complements):

	A	A′
B	P(A∩B)	P(A′∩B)
B′	P(A∩B′)	P(A′∩B′)

REMEMBER

All the probabilities in a 2×2 contingency table should sum to 1 because you are showing all the combinations of what could happen, which are represented by the $2 \times 2 = 4$ joint probabilities.

EXAMPLE

Q. Suppose that a survey shows that 50 percent of college football fans who go to a game will buy food and not merchandise, 20 percent will buy merchandise and not food, and 15 percent will buy both. What percentage will not buy food or merchandise?

A. The answer is 0.15. You have $P(F \cap M') = 0.50$, $P(F' \cap M) = 0.20$, and $P(F \cap M) = 0.15$. These pieces fit directly into three of the four cells of the table. You want $P(F' \cap M')$, which is the remaining cell. You can solve this problem by subtracting the sum of all the other cells from 1: $P(F' \cap M') = 1 - (0.50 + 0.20 + 0.15) = 0.15$. The contingency table is shown here:

	F	F′
M	0.15	0.20
M′	0.50	$1 - (0.50 + 0.20 + 0.15) = 0.15$

 Suppose that Petrie is shopping at two stores, Big Store and Big Deal. Petrie buys 5 percent of his items at Big Store but not Big Deal, 10 percent at Big Deal but not Big Store, and 35 percent at neither store. How much does Petrie spend at both stores?

2 Suppose that Geoff shops for groceries mainly at two stores, Savon and Costless. Geoff buys 20 percent of his items at both stores, 25 percent at Savon but not Costless, and 30 percent at Costless but not Savon. How often does Geoff buy groceries somewhere else besides these two stores?

 3 Suppose that Geoff shops for groceries mainly at two stores, Savon and Costless. Geoff buys 30 percent of his items at both stores, 25 percent at Savon but not Costless, and 30 percent at Costless but not Savon. How much does Geoff buy at Savon or Costless (or both)?

Finding and Interpreting Marginal Probabilities Using a Contingency Table

Contingency tables are useful when you are given or can find the intersection probabilities that go into the four cells. Once you have the table filled in, you can easily find the totals in the rows and columns by summing across or down to get the *marginal probabilities.*

 They are called marginal probabilities because they are out in the margins of the table.

TIP In the following table, you can see that when you sum across row 1, you get the probability of A and B plus the probability of A' and B, which sums to P(B). Similarly, the second row sums to $P(B')$. And those two together sum to 1.

You can also see that as you sum down column 1, you get the probability of A with B, and the probability of A without B; this sums to the marginal probability P(A). Similarly, column 2 sums to $P(A')$.

When you are summing down a column or across a row to get a marginal probability, you are using the Law of Total Probability. $P(A \cap B) + P(A' \cap B)$ is the same as $P(A)P(B|A) + P(A')P(B|A')$ using the multiplication rule, and is the same as pathway 1 + pathway 3 on the tree diagram in Chapter 3.

	A	A'	Totals
B	$P(A \cap B)$	$P(A' \cap B)$	$P(B)$
B'	$P(A \cap B')$	$P(A' \cap B')$	$P(B')$
Totals	$P(A)$	$P(A')$	1

Q. Lenny is flying home for the weekend on either airline A or B, and the plane will be on time or not on time with certain probabilities. The joint (intersection) probabilities are listed in the four cells of the following table. Fill in the four marginal probabilities and interpret.

A. See the marginal probabilities found and labeled in the last row and column of the following table. $P(A) = 0.60$ = the probability of flying on airline A; $P(B) = 0.40$ = the probability of flying on airline B. $P(OT) = 0.86$ = the probability of the plane being on time; and $P(OT) = 0.14$ = the probability of being late (not on time).

	On Time (OT)	Not On Time (OT)'	Totals
A	0.54	0.06	$P(A) = 0.54 + 0.06 = 0.60$
B	0.32	0.08	$P(B) = 0.32 + 0.08 = 0.40$
Totals	$P(OT) = 0.54 + 0.32 = 0.86$	$P(OT)' = .06 + .08 = 0.14$	1

 4 Boris teaches a probability class. Sixty percent of his students in the class are from the statistics department and are women; 5 percent of his students are from the math department and are women; 20 percent are from the statistics department and are men; and the rest are from the math department and are men. What percentage of Boris's students are from the math department?

Atalie is a photographer. They spend 50 percent of their time taking pictures and using the computer to do it; they spend 30 percent of their time editing and using the computer; they spend 15 percent of their time taking pictures without using the computer; and the rest of their time is spent editing without using the computer. What total percentage of Atalie's time is spent *not* working on the computer? Create a table.

 A hotel has two restaurants, A and B. Forty-eight percent of the customers eat at restaurant A and are satisfied, and 36 percent eat at restaurant B and are satisfied. What's the total percentage of customers who are *not* satisfied? Create a table. Fill in the information you are given and calculate what is left.

Hint: You don't need to — nor can you — fill in the entire table.

Finding and Interpreting Conditional Probabilities Using a Contingency Table

Conditional probabilities are found by taking intersection probabilities divided by marginal probabilities. Recall from Chapter 2, the formulas are $P(A|B) = \dfrac{P(A \cap B)}{P(B)}$ for the probability of A given B, and $P(B|A) = \dfrac{P(B \cap A)}{P(A)}$ for the probability of B given A. While conditional probabilities themselves are not found on a contingency table, the parts that go into their formulas are — the intersection goes on the top, and the marginal of what is "given" goes on the bottom.

WARNING

Never put conditional probabilities inside a contingency table. The cells inside a contingency table are for intersection (joint) probabilities only, and the margins are for marginal probabilities only.

To find the conditional probability of A|B, you do the following:

1. Find the intersection probability for A and B.

2. Divide by the marginal probability of B (what you are given).

To find the conditional probability of B|A, you do the following:

1. Find the intersection probability for B and A (which is the same as the one for A and B).

2. Divide by the marginal probability of A (what you are given).

EXAMPLE

Q. Suppose that a hotel has two restaurants: A and B. Customers are either satisfied (S) or not satisfied (NS) at these restaurants. The probabilities for the accompanying 2×2 contingency table are shown in the following table. What is the probability that a customer is satisfied given that the customer ate at Restaurant A? What is the probability that a customer is satisfied given that the customer ate at Restaurant B?

	S	NS	Totals
A	0.48	0.12	0.48 + 0.12 = 0.60
B	0.36	0.04	0.36 + 0.04 = 0.40
Totals	0.48 + 0.36 = 0.84	0.12 + 0.04 = 0.16	1

A. $P(S|A) = \dfrac{P(S \cap A)}{P(A)} = \dfrac{0.48}{0.60} = 0.80$ and $P(S|B) = \dfrac{P(S \cap B)}{P(B)} = \dfrac{0.36}{0.40} = 0.90.$

---

7   Boris teaches a probability class. Some of the students are from the statistics department, and some are from the math department. Gender is noted as well. The joint probabilities are listed in the following table. If a student is from the statistics department, what is the probability that the student is male?

	F (Female)	M (Male)	Totals
S (Statistics)	0.36	0.24	
M (Math)	0.16	0.24	
Totals			

Atalie is a photographer. When working, they spend some of their time taking pictures and the rest of their time editing pictures. Atalie uses the computer for a certain percentage of the time for each job. The known probabilities are listed in the following table. What percentage of Atalie's time is spent working on the computer if they are editing photos?

	Computer (C)	No Computer (NC)	Totals
Taking Photos	0.05	0.45	0.50
Editing Photos	0.48	0.02	0.50
Totals	0.53	0.47	1

 A vendor sells mangos and dragon fruit. At the end of the day, some of the mangos and dragon fruit that did not sell had to be thrown out. The probabilities are shown in the following table. M = mango; DF = dragon fruit; TO = thrown out; NTO = not thrown out. If a randomly selected fruit is a mango, what's the chance it was not thrown out?

	TO	NTO	Totals
M	0.05	0.50	
DF	0.03	0.42	
Totals			

Checking for Independence of Two Events in a 2 x 2 Table

Two events are independent in a 2×2 contingency table if any one of three scenarios is true:

» $P(A|B) = P(A)$

» $P(A|B) = P(A|B')$

» $P(A \cap B) = P(A)P(B)$

Basically, A and B are independent if knowing whether B occurred doesn't affect the probability of A occurring, and vice versa.

For example, in a standard deck of 52 cards, if A = a card is red and B = a card is a 2, A and B are independent. You only have to show that one of the three situations is true. Let's look at all three here:

» $P(Red|2) = \frac{2}{4} = \frac{1}{2}$, and $P(Red) = \frac{26}{52} = \frac{1}{2}$. These are equal, so red and 2 are independent.

» $P(Red|2) = \frac{2}{4}$ and $P(Red|Not\ 2) = \frac{24}{48} = \frac{1}{2}$, so again they are independent.

» $P(Red\ and\ 2) = \frac{2}{52}$ and $P(Red)P(2) = (\frac{1}{2})(\frac{4}{52}) = \frac{2}{52}$, so red and 2 are independent.

EXAMPLE

Q. Lenny is flying home for the weekend on either airline A or B, and the plane will be on time or not on time with certain probabilities. The joint (intersection) probabilities are listed in the four cells of the following table. Given this information, are these two variables independent?

A. Let's check using the first scenario: Is $P(A|B) = P(A)$? Let A = airline A and B = on time. $P(A|On\ Time) = P(A\ and\ On\ Time)/P(On\ Time) = 0.54/0.86 = 0.63$; and $P(A) = 0.60$. These are not equal, so the variables are not independent.

	On Time (OT)	Not On Time (OT)'	Totals
A	0.54	0.06	0.60
B	0.32	0.08	0.40
Totals	0.86	0.14	1

10. A vendor sells mangos and dragon fruit. At the end of the day, some of the mangos and dragon fruit had to be thrown out. The probabilities are shown in the following table. M = mango; DF = dragon fruit; TO = thrown out; NTO = not thrown out. Is the fruit the vendor is selling independent of whether or not it gets thrown out?

	TO	NTO	Totals
M	0.05	0.50	0.55
DF	0.03	0.42	0.45
Totals	0.08	0.92	1

11. Deb teaches a statistics class. Some students are from the statistics department, and some are from the math department. Gender is noted as well. The joint probabilities are listed in the following table. Given the information in the table, are gender and department independent?

	F (Female)	M (Male)	Totals
S (Statistics)	0.125	0.125	0.25
M (Math)	0.375	0.375	0.75
Totals	0.50	0.50	1

Suppose that Dallas teaches a probability class, and in this class, you know that gender and department are independent. Fill in the four cells of the following table to make the independence work.

Hint: Use one of the three scenarios of independence.

	F (Female)	M (Male)	Totals
S (Statistics)			0.35
M (Math)			0.65
Totals	0.45	0.55	1

 Tell whether the following statement is true or false: "You have to show all three scenarios for independence in order to say that A and B are independent."

14 Atalie is a photographer. When working, they spend some of their time taking pictures and the rest of their time editing pictures. Atalie uses the computer for a certain percentage of the time for each job. The known probabilities are listed in the following table. Is using the computer and taking pictures independent? Use the second scenario of independence to check this one.

	Computer (C)	No Computer (NC)	Totals
Taking Photos	0.05	0.45	0.50
Editing Photos	0.48	0.02	0.50
Totals	0.53	0.47	1

Solutions to Problems in Setting the Contingency Table with Probabilities

(1) The answer is 0.50. The information provided fills three of the four cells in the table, leaving only the upper-left cell. Because the cells sum to 1, the remaining cell has the probability of $P(BS \cap BD) = 1 - (0.05 + 0.10 + 0.35) = 050$.

	BD	NOT BD
BS	1 – (0.05 + 0.10 + 0.35) = 0.50	0.05
Not BS	0.10	0.35

(2) The answer is 0.25. The information provided fills three of the four cells in the table, leaving only the lower-right cell. Because the cells sum to 1, this remaining cell has the probability of $P(\text{Not } S \cap \text{Not } C) = 1 - (0.20 + 0.25 + 0.30) = 0.25$.

	C	NOT C
S	0.20	0.25
Not S	0.30	1 – (0.20 + 0.25 + 0.30) = 0.25

(3) The answer is 0.85. The information provided fills three of the four cells in the table, and those three cells constitute the answer to the problem. So, sum them up to get (0.30 + 0.25 + 0.30) = 0.85. (You can also use the addition rule, but it takes more work.)

	C	NOT C
S	0.30	0.25
Not S	0.30	

(4) The answer is 0.20. The information given in the table fills three of the four cells, and "the rest" is 1 minus the sum of those three cells: 1 – (0.60 + 0.20 + 0.05) = 0.15. Now you have the four cells of the table, so you can answer the question.

	W	M	Totals
Stat	0.60	0.20	0.80
Math	0.05	1 – (0.60 + 0.20 + 0.05) = 0.15	0.05 + 0.15 = 0.20
Totals	0.65	0.35	1

5) The answer is 0.20. The information given in the table fills three of the four cells, and "the rest" is 1 minus the sum of those three cells: $1 - (0.50 + 0.15 + 0.30) = 0.05$. Now you have the four cells of the table, so you can answer the question.

	Computer	No Computer	Totals
Taking Pictures	0.50	0.15	0.65
Editing	0.30	1 – (0.50 + 0.15 + 0.30) = 0.05	0.35
Totals	0.80	0.15 + 0.05 = 0.20	1

6) The answer is 0.16. Find P(S) and take the complement to get P(NS): $P(S) = P(A \cap S) + P(B \cap S) = 0.48 + 0.36 = 0.84$. So, $1 - 0.84 = 0.16$.

	S	NS	Totals
A	0.48		
B	0.36		
Totals	0.48+0.36 = 0.84	1– 0.84 = 0.16	1

7) The answer is 0.40. You want P(M|S) = P(M and S)/P(S) = 0.24/0.60 = 0.40.

	F (Female)	M (Male)	Totals
S (Statistics)	0.36	0.24	0.60
M (Math)	0.16	0.24	0.40
Totals	0.52	0.48	1

8) The answer is 0.96. You want P(Computer|Editing) = P(Computer and Editing)/P(Editing) = 0.48/0.50 = 0.96.

9) The answer is 0.91. You want P(NTO|M) = 0.50/0.55 = 0.91.

10) No, they are not independent. You can check P(TO|DF) = 0.03/0.45 = 0.067 and P(TO) = 0.08.

TIP

In this chapter, the values must be equal, not just close, to be independent.

	TO	NTO	Totals
M	0.05	0.50	0.55
DF	0.03	0.42	0.45
Totals	0.08	0.92	1

(11) Yes, they are independent. You can check $P(S \cap F) = 0.125$, and $P(S)P(F) = 0.25 * .50 = 0.125$. These are equal, so the events are independent.

	F (Female)	M (Male)	Totals
S (Statistics)	0.125	0.125	0.25
M (Math)	0.375	0.375	0.75
Totals	0.50	0.50	1

(12) Use $P(A \cap B) = P(A)P(B)$; the cell probability equals the row probability times the column probability for that cell.

	F (Female)	M (Male)	Totals
S (Statistics)	0.35 * 0.45 = .1575	0.35 * 0.55 = 0.1925	0.35
M (Math)	0.65 * 0.45 = 0.2925	0.65 * 0.55 = 0.3575	0.65
Totals	0.45	0.55	1

(13) False. You only have to show one of the three scenarios to show independence; the others are automatically true.

(14) No, they are not independent. You can check $P(\text{Taking} | \text{Computer}) = 0.05/0.53 = 0.9$, and $P(\text{Taking}) = 0.50$. These are not equal, so taking pictures and working on the computer are not independent.

	Computer (C)	No Computer (NC)	Totals
Taking	0.05	0.45	0.50
Editing	0.48	0.02	0.50
Totals	0.53	0.47	1

Chapter **5**

Unraveling Counting Rules

Probability is the number of ways a certain outcome can occur divided by the total number of possible outcomes of some random process, such as flipping a coin. For example, P(6) when rolling a die is ⅙ because there is one 6 on a die, and there are six total faces on the die. If you roll two dice and you want the total probability of getting a sum of 5, you first figure out the number of ways to get a sum of 5, then divide by the total number of outcomes of two dice (6 * 6 = 36). The number of ways to get a sum of 5 is four: You can get a 2 and a 3, which sum to 5, or you can get a 3 and a 2, which sum to 5 or you can get 4 and a 1 or a 1 and a 4, which sum to 5. Thus, the probability of getting a sum of 5 when rolling two dice is 4/36.

Figuring certain probabilities involves counting the number of ways to rearrange the desired outcomes when order makes a difference, such as the number of ways to seat five people in an aisle in the theater. Each new order counts. What's needed is a formula to determine the number of ways to rearrange the outcomes you are choosing. When order matters, we call this a *permutation*.

Figuring other probabilities involves counting the number of ways to rearrange the desired outcomes when order doesn't matter, such as pulling five names out of a hat containing 50 names. What matters is the five names, not the order in which they are chosen. What's needed is a formula to determine the number of ways to choose the outcomes. When order does not matter, we call this a *combination*.

In this chapter, you explore how to count the number of permutations and the number of combinations for various scenarios. These techniques are called *counting rules*. Counting rules are used in games of chance like poker, for example, or just to calculate the chance of some event occurring from scratch.

Pondering Permutations

The number of possible ways to rearrange k objects is $k*(k-1)*(k-2)*\ldots 3(2)(1)$. For example, the number of ways to rearrange six people in a row in the theater is $6*5*4*3*2*1 = 720$. That's because the first person has six places to choose from; the second person has five places to choose from; the next person has four; then three; then two; then the last person has one choice, and sits in the last open seat.

The mathematical shorthand for rearranging k items is k! (called "k-factorial"). For example, 3! equals three factorial, which is $3*2*1 = 6$, representing the six rearrangements of three items. This works for any k greater than or equal to 1. Note that 1! equals 1, and by convention, 0! also equals 1. (If you think about it, there is only one way to rearrange 0 items.)

If you want to first choose the k items from a total of n items without replacement (without putting an item back after it is chosen), and then rearrange them, you write this as $n\,P_k$ or $P_k^n = \dfrac{n!}{(n-k)!}$, or nPk, and you say this is a permutation of k items chosen from n items (without replacement). For example, if six people are going to the theater, and only four can sit together, you first choose the four people who are going to sit together, and then you rearrange them. The number of ways of doing this is $P_4^6 = \dfrac{6!}{(6-4)!} = \dfrac{6!}{2!} = \dfrac{6*5*4*3*2*1}{2*1} = 360$.

TIP The number on top or in front of the permutation notation is n, and the number in back or at the bottom of the permutation notation is k. It is always true that k is less than or equal to n. If you just want to rearrange all the items in the group, you have $P_n^n = \dfrac{n!}{(n-n)!} = \dfrac{n!}{0!} = \dfrac{n!}{1} = n!$.

The reason you divide by $(n-k)!$ is because in the numerator, you have the number of ways to rearrange all n of the values, and you only care about rearranging k of them (the ones you chose). So, you divide by the number of ways to rearrange the items you didn't select, which is $(n-k)$, which can be rearranged $(n-k)!$ times.

TIP With factorials, the numbers get large very quickly as more and more are multiplied together, so there is a shortcut for calculating permutations: Take the numerator and write it as $n*(n-1)*(n-2)\ldots(n-k)!$ and stop there, and then cancel out $(n-k)!$ with the denominator. For example, $P_2^{10} = \dfrac{10!}{(10-2)!} = \dfrac{10*9*8!}{8!} = 10*9 = 90$.

With a smaller example, you can list out the possible rearrangements. For example, suppose that you have four people — Anabav, Courtney, Xinyu, and Melissa — and you want to choose two to be the heads of a committee. One person will be the committee chair and the other the vice-chair. That means order matters, and you have to sample without replacement because one person can't hold both roles. How many ways can you fill the two roles, and what are the possible rearrangements?

In this example, you have $P_2^4 = \dfrac{4!}{(4-2)!} = \dfrac{4*3*2!}{2!} = 12$.

The 12 rearrangements are:

1. Anabav and Courtney
2. Anabav and Xinyu
3. Anabav and Melissa
4. Courtney and Xinyu
5. Courtney and Melissa
6. Xinyu and Melissa

But now you can switch the order of all six of these arrangements, since you would be switching who is the chair and who is the vice chair, so you also have:

7. Courtney and Anabav
8. Xinyu and Anabav
9. Melissa and Anabav
10. Xinyu and Courtney
11. Melissa and Courtney
12. Melissa and Xinyu

REMEMBER

Ask yourself, "Does the order of the selected item matter?" If so, use a permutation; if not, use a combination (see the next section).

Here's another example: How many ways can you rearrange k people sitting in a circle? The answer is $(k-1)!$. The reason is, you don't have a "starting point" to the circle, so you can start with k!, but then you divide out the number of circles that would be the same, and that would be k. You then get $(k-1)!$ For example, if three people — Fred, Arvin, and Sam, for example — are sitting in a circle, the only thing that matters is who is on the left and who is on the right of each person. If you tell Fred to sit in a specific spot in the circle, Arvin can be on Fred's left, and Sam can be at Fred's right. Or, Arvin could be on the right and Sam on the left. That's $\dfrac{3!}{3} = 2! = 2$.

Some problems are more complex and require you to put in more than one permutation. Suppose that you have four people sitting in a row in the theater: Bill, Bob, Mary, and Sue. What is the probability that Bill and Mary will end up sitting next to each other? First, find the denominator, which is the number of ways to rearrange the four people; this is 4! For the numerator, find the number of ways Bill and Mary could end up sitting together.

Think of Bill being to the left of Mary first (you have to be systematic about these problems). Bill has three choices, and Mary has to be on his right, so she has one choice. Then Bob and Sue can be rearranged in the remaining two spots, so the formula is:

$$P_1^3 * P_1^1 * P_2^2 = \frac{3!}{(3-1)!} * \frac{1!}{(1-1)!} * \frac{2!}{(2-2)!} = 3 * 1 * 2 = 6.$$

And here they are:

Bill, Mary, Bob, Sue

Bill, Mary, Sue, Bob

Bob, **Bill, Mary**, Sue

Sue, **Bill, Mary**, Bob

Bob, Sue, **Bill, Mary**

Sue, Bob, **Bill, Mary**

But you aren't finished yet! Notice that Mary and Bill can be swapped in every case. In the first case, you have **Bill, Mary**, Bob, Sue, but you could also have **Mary, Bill**, Sue, and Bob. That means you multiply all six results by $P_2^2 = \frac{2!}{(2-2)!} = 2$, which is the number of ways to rearrange the two people among themselves, bringing the total number of ways Bill and Mary can sit together to 12.

EXAMPLE

Q. Suppose that out of a group of ten people, you are choosing three of the people to win a prize. No one can win more than one prize, and each prize is different. How many different ways can you hand out the prizes? (Take a guess first.)

A. The answer is 720. That might be a much bigger number than your guess! This is a permutation problem because you are sampling without replacement, and the prizes are different, so order matters. And of the ten people, you pick three to win a prize, so the number of rearrangements is $P_3^{10} = \frac{10!}{(10-3)!} = \frac{10*9*8*7!}{7!} = 10*9*8 = 720$. One way to look at it is that the first prize can go to any of the ten people, but the second prize can only go to nine people since the first prize winner can't win again. Then the third prize can't go to either of the previous winners, so there are eight people left to give the prize to, and $10*9*8 = 720$.

1 Suppose that you have a group of five people and you are giving out two different prizes. No one can win more than one prize. How many ways are there to do this?

2 Suppose that you want to seat five people next to each other in the same row at the theater. How many ways are there to rearrange the five people in the seats?

3 Suppose that you want to seat five people next to each other in the same row at the theater. What is the probability that Bob and Bill sit on each end of the row, and the other three people sit in the middle?

4 Suppose that you have five seats around a table for Tommy, Ed, Oksana, Drew, and David. What is the total number of possible seating arrangements?

5 How many different batting orders are possible for a nine-player baseball team selected from a group of 15 players?

6 In how many ways can a president, vice-president, and secretary be chosen from a group of 12 club members?

7 How many four-digit personal identification numbers (PINs) can be made using the digits 0 through 9 if repeats are *not* allowed?

Figuring Permutation Probabilities

To figure probabilities involving permutations, you work with the numerator and denominator separately. The numerator is the number of ways to do the outcome you are interested in; the denominator is the total number of arrangements involved in the process.

For example, consider a horse race involving ten horses. How many different trifectas are there to bet on? (A trifecta occurs when you bet correctly on the first, second, and third place winners.) The denominator contains $_{10}P_3$ since you have ten horses and three are going to end up in first, second, and third place. Only one set of three is going to appear as the first, second, and third place winners, so the probability is $\dfrac{1}{_{10}P_3} = \dfrac{1}{10!/(10-3)!} = \dfrac{1}{10*9*8} = \dfrac{1}{720} = 0.0014$.

Q. Suppose that out of a group of ten people, you are choosing three of the people to win a prize. You are included in the group. No one can win more than one prize, and the prizes are different. What is the chance that you win one of the prizes?

A. The answer is $\frac{3}{720}$. This is a permutation probability problem because you are sampling without replacement, and the prizes are different, so order matters. In the denominator, you have the following: Of the ten people, you pick three to win a prize, so the number of rearrangements is $P_3^{10} = \dfrac{10!}{(10-3)!} = \dfrac{10*9*8*7!}{7!} = 10*9*8 = 720$. The numerator is the number of ways for you to win a prize, which is three, since there are three prizes. So, the probability is $\frac{3}{720} = 0.0042$.

8 Suppose that five horses enter a race. Each horse has a number on its saddle, 1 through 5. What's the chance that the horses finish in numerical order?

9 In a field of ten horses, what is the chance that the horse you chose wins?

10 Twelve members of a club select three officers at random. What is the chance that Benjamin, Olivia, and Kit are the president, vice-president, and the secretary, respectively?

11 Twelve members of a club select three officers at random. What is the chance that Benjamin, Olivia, and Kit are all officers?

12 You select three letters without replacement from the alphabet. What is the probability that the letters you choose spell "cat"?

13 You have four people at a table: Bob, Marion, Arushi, and Yipeng. They sit randomly around the table. What is the chance that Bob and Yipeng sit together?

Catching up on Combinations

While a permutation occurs when you sample without replacement and the order of the outcomes matters, a *combination* occurs when you sample without replacement and the order of the outcomes does not matter. In this case, your notation is $_nC_k$ or C_k^n or $\binom{n}{k}$ for a combination of k items chosen from n items. (You can also write it as nCk.)

Recall that the number of ways to pull off a permutation is counted using the formula $_nP_k = \frac{n!}{(n-k)!}$. You have n!, which is the total number of rearrangements of n items, and you divide by the number of rearrangements of the items you *don't* pick, which is $(n-k)!$ Because you don't pick them, you don't worry about their order. What's left is the order of the k items you did pick.

Now with combinations, you don't worry about the order of the items you don't pick (so you divide by $(n-k)!$), you also don't worry about the order of the items you *do* pick, which is k! (so you divide by k! as well). This divides the group into two subgroups: those you picked and those you didn't pick, and takes out the ordering of each group. The formula for a combination of k items taken from n items is $_nC_k = \frac{n!}{k!(n-k)!}$.

TIP

$_nC_k = \frac{n!}{k!(n-k)!}$ will be less than or equal to $_nP_k = \frac{n!}{(n-k)!}$ because you are dividing additionally by k! It's equal if $k=0$ or 1. So, unless you are picking 0 or 1 items, you'll find that the number of permutations is more than the number of combinations; that's because with permutations, order matters, and with combinations, order doesn't matter.

Let's work out a couple of combinations. For example, $_5C_2 = \frac{5!}{2!(5-2)!} = \frac{5*4*3!}{(2*1)3!} = \frac{20}{2} = 10$.

If this had been a permutation, you would have $_5P_2 = \frac{5!}{(5-2)!} = \frac{5*4*3!}{3!} = \frac{20}{1} = 20$.

Here's another example: $_4C_2 = \frac{4!}{2!(4-2)!} = \frac{4*3*2!}{(2*1)2!} = \frac{12}{2} = 6$. If you had the four letters A, B, C, D, and you want to choose two of those letters as a combination, you would get AB, AC, AD, BC, BD, CD, which makes six combinations. You would not further rearrange each combination in order, like you would a permutation. The permutation of $_4P_2 = \frac{4!}{(4-2)!} = \frac{4*3*2!}{2!} = \frac{12}{1} = 12$ would be double because you would have AB or BA, AC or CA, AD or DA, BC or CB, BD or DB, and CD or DC.

Note that the results are the same when $k=0$ or 1. $_4C_0 = \frac{4!}{0!(4-0)!} = \frac{4!}{(0!)4!} = \frac{1}{1(1)} = 1$, which is the same as $_4P_0 = \frac{4!}{(4-0)!} = \frac{4!}{4!} = \frac{1}{1} = 1$. The same is true for $_4C_1 = \frac{4!}{1!(4-1)!} = \frac{4*3!}{(1)3!} = \frac{4}{1} = 4$, which is the same as $_4P_1 = \frac{4!}{(4-1)!} = \frac{4*3!}{3!} = \frac{4}{1} = 4$. (When you choose 0 or 1 item, there is only one way to order it.)

REMEMBER

The main question when doing problems involving permutations and combinations is, "Does the order of the selected items matter?" If yes, do a permutation. If no, do a combination.

EXAMPLE

Q. How many ways are there to choose three people out of a group of six to win a prize if the prizes are all the same?

A. This is a combination. Because the prizes are all the same, order doesn't matter. The answer is

$$_6C_3 = \frac{6!}{3!(6-3)!} = \frac{6*5*4*3!}{(3*2*1)3!} = \frac{5*4}{1} = 20.$$

14 How many ways are there for three horses in a race of seven to "win, place, or show?" (That means they come in one of the first three places, and their order doesn't matter.)

15 A local pizza shop offers 12 different toppings. How many different three-topping pizzas can be ordered?

16 How many different five-card hands can be dealt from a standard 52-card deck?

17 A committee of four people is to be selected from a group of ten employees. How many different committees can be formed?

18 How many ways are there for ten people at a theater to sit in two rows of five?

19 A math test includes six questions. You can pick any three questions to do. How many sets of questions are possible?

20 How many ways are there to choose five items for your salad when the salad bar offers 15 items?

Probabilities Involving Combinations

Poker hands from a 52-card deck are a common way to think about combination probabilities. In the numerator, you figure out how many ways there are to get the type of hand you want, and the denominator is the total number of hands:

$$_{52}C_5 = \frac{52!}{5!(52-5)!} = \frac{52*51*50*49*48*47!}{(5*4*3*2*1)47!} = \frac{52*51*50*49*48}{120} = 2{,}598{,}960$$

REMEMBER

A standard deck includes 52 cards, with four different suits (clubs, spades, diamonds, and hearts) and 13 different denominations: 2, 3, 4, 5, 6, 7, 8, 9, 10, jack, queen, king, and ace. For example, there is a 2 of hearts, a 2 of clubs, a 2 of diamonds, and a 2 of spades, and so on, and $4*13 = 52$ total cards. A five-card hand means you choose five cards at random without replacement from the deck all at once (so order doesn't matter).

Q. What is the probability of getting a flush in a five-card poker hand? (A flush is when all cards are of the same suit.)

EXAMPLE

A. The denominator is $_{52}C_5 = \binom{52}{5} = 2{,}598{,}960$. The numerator is the total number of flushes. It is helpful to look at this problem in steps. First, you pick the suit, of which there are four, then once you pick the suit (say clubs), you pick five cards from the 13 cards of that suit. The number of ways to do this is $\binom{13}{5}$. Because both of these conditions have to be met, you multiply the number of ways to do them, so the total number of flushes is $\binom{4}{1}*\binom{13}{5}$. The final probability is

$$\left[\binom{4}{1}\binom{13}{5}\right]/\binom{52}{5} = \left(\frac{4!}{1!(4-1)!}\right)\left(\frac{13!}{5!(13-5)!}\right)/2{,}598{,}960 =$$

$$\left(\frac{4(3!)}{1(3!)}*\frac{13(12)(11)(10)(9)(8!)}{5(4)(3)(2)(1)(8!)}\right)/2{,}598{,}960 = \left(4*\frac{13(12)(11)(10)(9)}{120}\right)/2{,}598{,}960 =$$

$$5{,}148/2{,}598{,}960 = 0.00198.$$

---

21 Suppose that a player enters a pick-three drawing in which they pick three different numbers from 0 to 9. Order doesn't matter. What's the chance the player wins by matching all three numbers chosen?

22 Suppose that a player buys a lottery ticket for which they pick six different numbers from 1 to 49. What is the probability that the player wins by matching all six numbers chosen?

23 A group has eight men and seven women. A committee of five people is selected at random. What is the probability that the committee consists of exactly three men and two women?

24 What is the probability of being dealt a five-card poker hand that contains exactly four aces?

25 A box contains ten items, three of which are defective. If a quality control inspector randomly selects four items from the box for testing, what is the probability that at least one of the selected items is defective?

Solutions to Problems in Unraveling Counting Rules

(1) Order matters because the prizes are different; therefore, it's a permutation. The equation is $_5P_2 = \dfrac{5!}{(5-2)!} = \dfrac{5*4*3!}{3!} = 20$.

(2) Order matters, so it's a permutation. You have five people in five seats, so the equation is $_5P_5 = \dfrac{5!}{(5-5)!} = \dfrac{5*4*3*2*1}{1} = 120$.

(3) Let X represent the other people besides Bill and Bob. You can have Bob, X, X, X, Bill, or Bill, X, X, X, Bob. There are $_3P_3 = \dfrac{3!}{(3-3)!} = \dfrac{3*2*1}{1} = 6$ ways to arrange the people in the middle, and you multiply by 2 because of the two ways Bill and Bob can be seated. So you have $2*6 = 12$.

(4) You have five people around a circle, so there are $_4P_4 = \dfrac{4!}{(4-4)!} = \dfrac{4*3*2*1}{1} = 24$ ways to rearrange them. Another way to look at this is that there are five people, so you start with 5! Then five circles are the same since there is no "starting point," so you divide by 5 to get $(5-1)! = 4!$.

(5) $_{15}P_9 = \dfrac{15!}{(15-9)!} = \dfrac{15*14*13*12*11*10*9*8*7*6!}{6!} = 1,816,214,400$. This is a huge number, probably way bigger than you thought. Numbers get large quickly when you multiply them together. (That's why the mega millions are so hard to win.)

(6) Order matters because of the different offices, so the equation is $_{12}P_3 = \dfrac{12!}{(12-3)!} = \dfrac{12*11*10*9!}{9!} = 1,320$.

(7) Repeats are not allowed, so order matters. The equation is $_{10}P_4 = \dfrac{10!}{(10-4)!} = \dfrac{10*9*8*7*6!}{6!} = 5,040$.

(8) Order matters, and you have only one way they can finish in numerical order, so the equation is $1/(_5P_5) = 1/\left(\dfrac{5!}{(5-5)!}\right) = 1/\left(\dfrac{5*4*3*2*1}{0!}\right) = \dfrac{1}{120} = 0.0083$.

(9) There is only one way your horse can win in a race between ten horses, so the answer is $\dfrac{1}{10} = 0.10$.

Or you can look at the problem this way: You divide by the ways the other nine horses can finish because they don't matter, and you get: $1/(_{10}P_1) = 1/\left(\dfrac{10!}{(10-1)!}\right) = 1/\left(\dfrac{10!}{9!}\right) = 1/\left(\dfrac{10}{1}\right) = \dfrac{1}{10} = 0.10$.

(10) There is only one way these three can win the three offices, respectively. The denominator is the number of ways to choose three people from 12 where order matters, so the equation is $1/_{12}P_3 = 1/\left(\dfrac{12!}{(12-3)!}\right) = 1/\left(\dfrac{12*11*10*9!}{9!}\right) = \dfrac{1}{1,320} = 0.0008$.

(11) There are $_3P_3 = \dfrac{3!}{(3-3)!} = 3! = 6$ ways to rearrange Kit, Benjamin, and Olivia into the three offices. The denominator is $_{12}P_3 = \left(\dfrac{12!}{(12-3)!} \right) = \left(\dfrac{12*11*10*9!}{9!} \right) = 1,320$. So the answer is $^6/_{1,320} = 0.0045$.

(12) There is only one way to spell cat out of $_{26}P_3 = \dfrac{26!}{(26-3)!} = \dfrac{26*25*24*23!}{23!} = 15,600$ ways to draw three letters from the alphabet. (Order matters.) So the equation is $^1/_{15,600} = 0.00006$.

(13) The answer is 0.67. The answer is easiest found using factorials. You have four people at a table: Bob, Marion, Arushi, and Yipeng. They sit randomly around the table. What is the chance that Bob and Yipeng sit together? There are four ways for Bob and Yipeng to sit together: two ways in each scenario. First, Bob is on the left, and Yipeng is on the right, and the other two people are rearranged in any way (which gives you two options); or Bob is on the right, and Yipeng is on the left, and the other two people are rearranged in any way (which gives you two options). So you have 1(1)(2!) for the first scenario, then (1)(1)(2!) for the second scenario, divided by 3!, which is the number of ways to rearrange four people around a table. So you get $^{(2+2)}/_6 = ^2/_3 = 0.67$. (The only way they don't sit together is if they are on opposite sides of the table, and the other two people are rearranged in any way.)

(14) You have three horses in a race of 7 to "win, place, or show?" Their order doesn't matter so it's a combination: $C_3^7 = \dfrac{7!}{3!(7-3)!} = \dfrac{7*6*5*4!}{6(4!)} = 35$.

(15) The order of pizza toppings doesn't matter, so the equation is $_{12}C_3 = \dfrac{12!}{3!(12-3)!} = \dfrac{12*11*10*9!}{6(9!)} = 220$.

(16) The order of the five cards doesn't matter, so the equation is $_{52}C_5 = \dfrac{52!}{5!(52-5)!} = \dfrac{52*51*50*49*48*47!}{120(47!)} = 2,598,960$.

(17) The order doesn't matter since the committee doesn't have separate positions, so the equation is $_{10}C_4 = \dfrac{10!}{4!(10-4)!} = \dfrac{10*9*8*7*6!}{24(6!)} = 210$.

(18) To solve this problem, you first pick five people from ten, and the order doesn't matter. The last five people have to sit in the second row; there are no choices about that. So the equation is $_{10}C_5 = \dfrac{10!}{5!(10-5)!} = \dfrac{10*9*8*7*6*5!}{120(5!)} = 252$. Or you could think of the problem this way: you first pick five from ten people with 252 ways, then you pick the last five people in one way since $_5C_5 = \dfrac{5!}{5!(5-5)!} = \dfrac{5!}{5!(1)} = 1$.

(19) There are six total questions, and you choose three to do. The number of ways is $_6C_3 = \dfrac{6!}{3!(6-3)!} = \dfrac{6!}{3!(3!)} = \dfrac{6*5*4*3!}{6*3!} = 20$.

(20) There are 15 items and you choose five to build your salad. The number of ways is $_{15}C_5 = \dfrac{15!}{5!(15-5)!} = \dfrac{15*14*13*12*11*10!}{120(10!)} = \dfrac{360,360}{120} = 3,003$.

(21) The ten numbers to pick from are picked without replacement, and order doesn't matter. The chance of winning is having the three numbers chosen. The denominator is $_{10}C_3 = \dfrac{10!}{3!(10-3)!} = \dfrac{10*9*8*7!}{6(7!)} = \dfrac{10*9*8}{6} = 120$. You have three numbers that have to match, and there is only one set of three that wins, so the probability is $\frac{1}{120} = 0.0083$.

(22) There is one winner, so "1" is the numerator. The denominator is C_6^{49}.

$$1/\left(_{49}C_6\right) = \dfrac{\frac{1}{49!}}{\frac{1}{6!43!}} = \dfrac{\frac{1}{49*48*47*46*45*44*43!}}{\frac{720*43!}{}} = \dfrac{\frac{1}{49*48*47*46*45*44*43!}}{720*43!} =$$

$$\dfrac{\frac{1}{49*48*47*46*45*44}}{720} = \dfrac{1}{13,983,816} = .000000072.$$

(23) There are $8+7 = 15$ people to choose from. So the denominator is $_{15}C_5 = \dfrac{15!}{5!(15-5)!} =$ $\dfrac{15*14*13*12*11*10!}{120(10!)} = \dfrac{15*14*13*12*11}{120} = 3,003$. The numerator is asking for three men out of eight chosen, and two women out of seven chosen, so the numerator is

$$_8C_3 *_7C_2 = \dfrac{8!}{3!(8-3)!} * \dfrac{7!}{2!(7-2)!} = \dfrac{8*7*6*5!}{(3*2*1)5!} * \dfrac{7*6*5!}{(2*1)5!} = 56 * 21 = 1,176.$$

That means the probability is $\frac{1,176}{3,003} = 0.3916$.

(24) There are 48 ways to choose four aces in a five-card poker hand. First, you choose all four of the four aces, and there is only one way to do that. Then, of the 48 remaining cards, you choose one, so the equation is $_4C_4 *_{48}C_1 = \dfrac{4!}{4!(4-4)!} * \dfrac{48!}{1!(48-1)!} = 1 * \dfrac{48*47!}{1(47)!} = 1 * 48 = 48$.

The denominator is $_{52}C_5 = \dfrac{52!}{5!(47)!} = \dfrac{52*51*50*49*48*47!}{120(47!)} = 2,598,960$. So the answer is $\frac{48}{2,598,960} = 0.000018$.

(25) For this problem, it's easier to think about the complement (or opposite) of "at least one," which is "none of the selected items is defective." So you find 1 – P(no defective items are chosen). The probability of 0 defective items means all are okay, so the quality control inspector must reach into the box containing ten items and pull out four of the $(10-3) = 7$ non-defective items. The denominator is 10C4. The numerator is 7C4. So the final answer is

$$1 - \dfrac{_7C_4}{_{10}C_4} = 1 - \dfrac{\frac{7!}{(4!(7-4)!)}}{\frac{10!}{(4!(10-4)!)}} = 1 - \dfrac{\frac{7*6*5*4!}{(4!*6)}}{\frac{10*9*8*7*6!}{(24(6)!)}} = 1 - \dfrac{\frac{7*6*5}{(6)}}{\frac{10*9*8*7}{(24)}} = 1 - \dfrac{35}{210} = 1 - 0.1667 = 0.8333.$$

Chapter **6**

Against All Odds: Probability in Gaming

When playing games of chance, it is important to keep in mind the laws of statistics and probability and understand the difference between odds, probability, and expected value.

A *probability* is a single number representing the proportion of favorable outcomes relative to the total number of possible outcomes. It is a long-term value. For example, when rolling a standard six-sided die, the probability of rolling a 4 is ⅙. It doesn't mean that if you roll a die six times, you'll get a 4 one time; it means that in the long term, if you roll infinitely many times, you'll get a 4 one-sixth of the time.

Odds express the likelihood of an event as a ratio of favorable outcomes to unfavorable outcomes. There are two types: odds in favor and odds against. The formula for odds in favor is favorable outcomes to (:) unfavorable outcomes. The formula for odds against is unfavorable outcomes to (:) favorable outcomes. For example, when rolling a standard six-sided die, the odds in favor of rolling a 4 are 1:5 (one favorable outcome to five unfavorable outcomes: 1, 2, 3, 5, 6). The odds against rolling a 4 are 5:1. It depends on the scenario you are involved in as to whether the odds are for or against. Odds compare the chances of something happening versus something *not* happening.

The *expected value* is the overall long-term average amount you'll win (if positive) or lose (if negative) per play. It takes into account the amount of money you put in per play. In any casino game, the expected value is generally going to be negative, even if it's just a little bit negative per play, because the casinos have to make money somehow.

In this chapter, I go over the basics of some of the more popular casino games. For full details on these and other casino games, check out *Probability For Dummies*, which is also published by Wiley and written by yours truly.

Playing the Lottery

The major lotteries available in the United States are generally split into two categories: games offered across many states and individual state-specific games. The two largest and most widely known lotteries are Powerball and Mega Millions, famous for their large, constantly growing jackpots that can get absolutely huge due to the large number of tickets sold. While the odds and payout systems of these lotteries are beyond the scope of this book, I do want to show you how many possible combinations there are with these lotteries, and how they can get much bigger with just a small change.

The Mega Millions lottery (at the time of this writing) selects five random numbers from 1 to 70 (printed on white balls) and one random number from 1 to 24 (called the Mega Ball). How many possible combinations of five balls are there? Using the counting rules from Chapter 5, the number of ways to choose five numbers without replacement from 70 numbers is the following:

$$_{70}C_5 = \frac{70!}{5!\,(70-5)!} = \frac{70*69*68*67*66*65!}{120(65!)} = \frac{70*69*68*67*66}{120} = 12,103,014.$$

Then, this number is multiplied by 24 because you also have to hit the Mega Ball to get the grand prize, and this brings the total number of possible combinations on a ticket to $12,103,014 * 24 = 290,472,336$.

TIP

Notice how big the number jumped when you added the Mega Ball, which has (only) 24 possible values. When you multiply numbers together, the results get big very quickly.

Suppose that we move from 70 numbers to 71 to choose from. And we move the Mega Ball from 24 possibilities to 25 possibilities. How much bigger do the possibilities get? With these changes, we have $_{71}C_5 = \frac{71!}{5!\,(71-5)!} = \frac{71*70*69*68*67*66!}{120(66!)} = \frac{71*70*69*68*67}{120} = 13,019,909$ possible combinations. That's a big difference from 12,103,014. Then, we multiply by 25 to get 325,497,725 combinations. In the end, it's important to check how many numbers you are choosing from before you play. It makes a difference. And the number of possible values that the balls can take on keeps increasing over time.

TIP

Another thing to think about when you are playing the big lotteries is to pick a set of numbers that no one else will pick, so then when the pot gets over $1 billion, you have a lower chance of having to split it with anyone. Another tactic is to avoid commonly chosen numbers (like the numbers between 1 and 31, which represent dates). That's the only edge you can get for yourself. For example, pick a sequence like 41, 42, 43, 44, 45 for the white balls and 1 for the Mega Ball. You might say, "Yeah, but that'll never win!" And I'll say, it has the exact same chance as any other combination out there, which puts in perspective how small a chance you have of winning in the first place.

Another idea is to play smaller statewide lotteries, which have smaller jackpots but increased chances of winning.

WARNING

Stick to a budget. Unless you are entering a pool where everyone pays a little, and you have a lot of people, it's better not to buy more tickets just to increase your chances a tiny amount. You'd have to buy *a lot* more to increase your chances enough to make it worthwhile, and that could bust your budget quickly.

EXAMPLE

Q. Tell whether the following statement is true or false: "The lotteries allow for repeats of numbers."

A. False. Lotteries pull numbers without replacement.

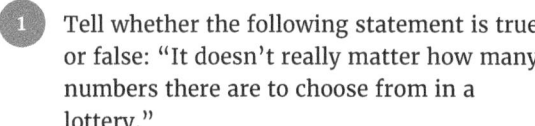

Tell whether the following statement is true or false: "It doesn't really matter how many numbers there are to choose from in a lottery."

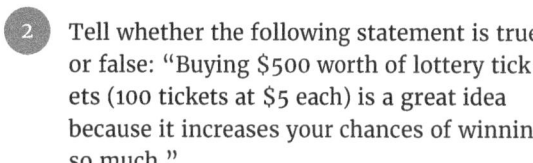

Tell whether the following statement is true or false: "Buying $500 worth of lottery tickets (100 tickets at $5 each) is a great idea because it increases your chances of winning so much."

Tell whether the following statement is true or false: "A set of numbers like 35, 36, 37, 38, 39, and 1 could never win, so I shouldn't bet on them."

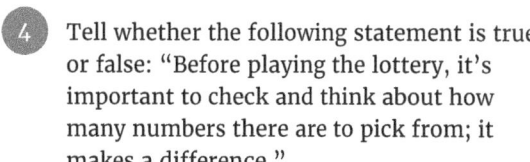

Tell whether the following statement is true or false: "Before playing the lottery, it's important to check and think about how many numbers there are to pick from; it makes a difference."

Betting on Blackjack

Very basically speaking, blackjack, or "21," is a card game in which you try to get as high a hand as you can without going over 21. You're playing against a dealer, who is also trying to win. Everyone starts with two cards. Each round, you can take another card ("hit") or not ("stay" or "stand"), and you can continue to take cards until you go over 21 ("bust") or you decide to stop. The dealer has a fixed strategy; typically they continue to take cards until they get 17, 18, 19, 20, or 21, then they stop. You try to beat their total by using your own strategy, and your whole goal is to see who gets closest to or equal to 21 without going over. You take your turn first, then the dealer takes their turn. A bet is placed before the cards are dealt. Ties result in a "push," which means you neither win nor lose the hand.

You can split your hand into two hands if you get two cards of the same value. And you can "double down" by placing an additional bet and getting one additional card only, after you see what your two initial cards are. The full rules are more complex, but this is the gist of it.

When playing blackjack, the question is, what is your strategy going to be? Some people stop when they first get over 12, so there is no chance of them going over 21 and busting (very conservative). Some people continue to take cards until they bust or get 20 or 21 (very risky). Some people make decisions based on what the cards are looking like as they come out of the combined decks (known as the "shoe"). This follows the "card-clumping" strategy. (Dealers use multiple decks all shuffled together to avoid card counting.)

One strategy is called the Basic Strategy, which works fairly well. The Basic Strategy was first developed by four U.S. Army engineers and published in 1956. It was later refined by mathematicians. It is a set of probability-based rules that tells a player what to do (hit, stand, double down, or split) for every possible hand they have against every possible dealer's card to give them the best chance of beating the house (dealer).

TIP

If you want to see the Basic Strategy, calculate your odds for a situation, and/or learn tips and techniques from some well-known blackjack players, check out the Wizard of Odds website at `https://wizardofodds.com/games/blackjack`.

You can get a little bit of an idea of what is going on if you assume that the shoe only has one deck, and you don't take into account what cards have already been played. (Things get really complicated, really fast in the real game.) For example, what's the chance that your first two cards will add up to 20? That's almost as close to 21 as you can get, and the Basic Strategy tells you to stop there. What is the chance of getting 20 on two cards? Using the counting rules from Chapter 5, you know the following: 10, jack, queen, and king are each worth 10. There are four suits, and so there are four 10s, four jacks, four queens, and four kings, for a total of 16 cards in a deck that are worth 10 points. The number of ways to get two 10-point cards when being dealt

two cards (assuming one 52-card deck), is $_{16}C_2 = \dfrac{16!}{2!\,(16-2)\,!} = \dfrac{16(15)(14!)}{2!\,(14!)} = 120$ (numerator). And the number of ways to choose two cards (assuming one 52-card deck, which is also called a shoe) is $_{52}C_2 = \dfrac{52!}{2!\,(52-2)\,!} = \dfrac{52*51*50!}{2*50!} = 1{,}326$ (denominator). So the probability of getting 20 points on the first two cards is $^{120}\!/_{1{,}326}$ or 9.05 percent. In this case, you stay, and hope the dealer doesn't whip out an ace and a 10-point card.

What's the chance of that happening? First, you need an ace, and there are $_4C_1 = \frac{4!}{1!\,(4-1)!} = \frac{4*3!}{1*3!} = 4$ ways to get that. Then you need a 10-point card, and there are $_{16}C_1 = \frac{16!}{1!\,(16-1)!} = \frac{16*15!}{1*15!} = 16$ ways to do that. Putting these together, you get $4*16 = 64$, out of 1,326 sets of two cards, which gives you a probability of $^{64}/_{1,326} = 4.8$ percent. In that case, the dealer wins. (For more specific probabilities in real-game situations, see the Wizard of Odds website at https://wizardofodds.com.)

EXAMPLE

Q. Tell whether the following statement is true or false: "Blackjack is a card game in which the goal is to hit 21 or over before the dealer does."

A. False. The goal is to get as close to 21 as possible without going over and beat the dealer.

---

5 How many two-card hands are possible to get dealt in blackjack?

6 What is the probability of getting a two-card blackjack?

7 What's the number of possible two ten-point cards, totaling 20 points?

8 What's the chance of getting 20 points on two cards you're dealt?

Reviewing a Bit about Craps

Craps is a fast-moving game of nonstop excitement and emotional highs and lows. Here is the very, very basic idea of craps (there are a lot of other things you can do on the table, but this is the gist of it): Players place their bets on the pass line and take turns "shooting" (rolling two dice). The shooter rolls the dice for the first, or "come-out" roll. A roll that totals 7 or 11 wins automatically, while a roll that totals 2, 3, or 12 loses automatically. If any other total is rolled besides a 7, 11, 2, 3, or 12, that number becomes the "point," and the shooter continues to roll until they either roll the point again (in which case you win) or they roll a 7 (in which case you lose). Players make bets throughout the game between rolls.

Before you run to the nearest craps table at your local casino with armfuls of money, you need to learn about what can happen when you roll two dice. Here are the 6 * 6 = 36 outcomes that you can get:

1, 1	2, 1	3, 1	4, 1	5, 1	6, 1
1, 2	2, 2	3, 2	4, 2	5, 2	6, 2
1, 3	2, 3	3, 3	4, 3	5, 3	6, 3
1, 4	2, 4	3, 4	4, 4	5, 4	6, 4
1, 5	2, 5	3, 5	4, 5	5, 5	6, 5
1, 6	2, 6	3, 6	4, 6	5, 6	6, 6

The total of the two dice is what is called out after each roll and what is bet on. Table 6-1 outlines the various possible totals for two dice and their probabilities (along with some of the lingo):

TABLE 6-1   Totals, Outcomes, and Probabilities of Rolling Two Dice

Total	Outcomes	Probability
2	1,1	$\frac{1}{36}$
3	1,2 or 2,1	$\frac{2}{36}$
4	1,3; 3,1 (called "easy 4") 2,2 (called "hard 4")	$\frac{3}{36}$
5	1,4; 4,1; 2,3; 3,2	$\frac{4}{36}$
6	1,5; 5,1; 2,4; 4,2 ("easy 6") 3,3 ("hard 6")	$\frac{5}{36}$
7	1,6; 6,1; 2,5; 5,2; 3,4; 4,3	$\frac{6}{36}$ (highest probability)
8	2,6; 6,2; 3,5; 5,3 ("easy 8") 4,4 ("hard 8")	$\frac{5}{36}$
9	3,6; 6,3; 4,5; 5,4	$\frac{4}{36}$
10	4,6; 6,4 ("easy 10") 5,5 ("hard 10")	$\frac{3}{36}$
11	5,6; 6,5	$\frac{2}{36}$
12	6,6	$\frac{1}{36}$

You can bet on all kinds of outcomes, such as whether a "hard way" or an "easy way" will come up. For more information, see `https://wizardofodds.com/games/craps/basics`.

EXAMPLE

Q. Tell whether the following statement is true or false: "Craps is a dice game involving one die."

A. False. Craps is a dice game involving two dice.

---

9 What is the chance that you roll any kind of 8 (total) with two dice?

10 What is the most difficult outcome to get (total) when you roll two dice?

---

11 How many ways are there to roll an even number (total) on two dice?

Poking into Poker Hands

In basic terms, poker is a card game in which the object is to get the best combinations of five cards by keeping and replacing cards on your turn and making bets. During betting rounds, you can bet the current bet, raise the bet, call (which means stay), or fold (which means give up) to influence or leave the pot. The strongest hand at the end, or the player with no one left to bet against, wins. The cards are ranked from high to low: ace, king, queen, jack, 10, 9, 8, 7, 6, 5, 4, 3, 2, ace. (Ace can be high or low, but is usually high.)

You can get ten different types of poker hands, each with different values. From top (most valuable) to bottom (least valuable but still worth something), here are the ten types of five-card hands you can end up with in poker:

1. Royal Flush (all the same suit and all in a row starting with 10: 10, J, Q, K, A)

2. Straight Flush (all the same suit, five in a row, not a Royal Flush)

3. Four-of-a-Kind (four cards of one kind and one other card)

4. Full House (three cards of one same kind and two of a different same kind)

5. Flush (all the same suit but no pattern)

6. Straight (all in a row, but not the same suit)

7. Three-of-a-Kind (three cards of the same kind, and two different cards)

8. Two Pair (two different pairs of cards and a fifth card not related)

9. One Pair (one pair of cards and three cards not related)

10. High Card (the highest card in the hand wins)

You can use the counting rules of Chapter 5 to figure out the number of hands of each type that exist in a 52-card deck. In *Probability For Dummies*, I go over how many hands are available for each type for most of the ones on the list. Here, I go over a couple of hands and give you the results of the number of ways to get each hand in Table 6-2.

Suppose that you want to know the number of ways to get a Full House. For example, 3-3-3-2-2. Using the counting rules from Chapter 5, you first pick the two "kinds" that you need, so that's $_{13}C_2 = 78$. Then of the two kinds you picked (say 3s and 2s), you pick one to be the Three of a Kind and one to be the Two of a Kind. There are two ways to do that (meaning you could have had 3-3-3-2-2 or 2-2-2-3-3), so that's $_2C_1$ equals 2. Then of the four cards that are the Three of a Kind, you pick three of them, so that's $_4C_3$, which is 4. Finally, of the four cards that are the Two of a Kind, you pick two of them, so that's $_4C_2$, which is 6. Put them all together, and you get: $78 * 2 * 4 * 6 = 3{,}744$ ways to get a Full House.

To get a Straight Flush, you pick your starting card (the lowest number in the straight), and there are $_{10}C_1$ possible ways to do that for each suit. (You can start with ace, 2, 3, 4, 5, 6, 7, 8, 9, 10.) There are four suits, so you must choose one for the flush (all one suit) part, which makes $_4C_1$ ways. Then the other four cards are forced to be what they are (only one way total). For example, if you have a 4 of hearts, the next cards *must* be 5 of hearts, 6 of hearts, 7 of hearts, 8 of hearts, and 9 of hearts. Altogether, there are only $10 * 4 = 40$ straight flushes available in the deck.

TABLE 6-2

The Hierarchy of Types of Poker Hands and the Number of Hands Available for Each Type

Hand	Number of Ways
Royal Flush	4
Straight Flush	40 − 4 = 36 (not counting Royal Flushes)
Four of a Kind	624
Full House	3,744
Flush	5,148 − 40 = 5,108 (not counting Straight and/or Royal Flushes)
Straight	10,240 − 40 = 10,200 (not counting Straight and/or Royal Flushes)
Three of a Kind	54,912
Two Pair	123,552
One Pair	1,098,240
High Card	1,302,540
TOTAL	**2,598,960**

EXAMPLE

Q. How many poker hands are there in total? (How do you find it using counting rules?)

A. The total number of poker hands is

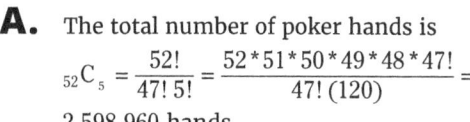

$${}_{52}C_5 = \frac{52!}{47!\,5!} = \frac{52 * 51 * 50 * 49 * 48 * 47!}{47!\,(120)} =$$

2,598,960 hands.

 12   What's the probability of a poker hand containing a Three of a Kind (but not a Full House)?

 13   How many Four of a Kinds are there?

14 How many Four of a Kinds are there that contain face cards only? (A face card is a J, Q, or K.)

15 How many Four of a Kinds are there that contain four aces only?

Taking a Spin with the Roulette Wheel

Roulette is one of the toughest games to win in the casino. It draws big crowds, and it gives the feeling you can be "on a roll," but that's not possible. Each spin of the wheel is independent of the rest, and that's something to remember if you decide to play it.

The American roulette wheel has 36 numbers (from 1 to 36) made up of 18 black numbers and 18 red numbers. So that the house has the edge — and it always does — there is one 0 and one 00 on the wheel, both green, yielding a total of 38 possible outcomes.

The betting layout consists of every individual number as well as a host of "outside" combinations of numbers. After the players make their bets, the dealer spins the wheel and drops a ball into it after a few seconds. The ball eventually lands in one of the numbered slots on the wheel, and the 38 numbers are equally likely to come up.

You can bet on many different things on a roulette wheel. One of the things you can bet on is the color of the slots the ball lands in (black or red). In this case, if a green comes up (0 and 00 are green), you lose; that's the house edge. A red or black bet pays even money, 1 to 1. That means you get your money back plus the amount of your bet. You can also bet odd or even, and it pays 1 to 1. You can bet on the lower half of the numbers (1 to 18) or the higher half of the numbers (19 to 36), and it pays 1 to 1. You can also bet on dozens, of which there are three: 1 to 12, 13 to 24, 25 to 36. This pays 2 to 1. That means you get your money back plus two times what you bet. You can also bet on columns of numbers, where the 36 numbers are divided into three different columns (one column is 1, 4, 7, 10, 13, 16, 19, 22, 25, 28, 31, 34), and you bet on which column will come up; this pays 2 to 1.

There are also straight bets, split bets, and street bets, which you can learn about at `https://wizardofodds.com`, or in *Probability For Dummies*.

WARNING

Sometimes people do win big on the roulette table, and if they quit while they were ahead, they could walk away happy. But you have to do that in order to walk away happy — quit while you are ahead. It is also important to quit if you get to a certain low point. Set those up in advance and stick to them.

TIP

When you play any casino game that involves betting (and which casino game doesn't?), you're playing against a house that has probability on its side. Studies will show (and probability theory bears this out) that if you play any game of chance long enough without stopping, you eventually lose everything, and possibly even more. This is called "Gambler's Ruin." Again, if you get a certain amount ahead (and set that ahead of time), quit when you get there. Or quit when you are in the hole a certain amount, whichever comes first.

EXAMPLE

Q. What's the chance of the ball landing on black in a roulette wheel?

A. There are 38 numbers. Two are green, and the rest are split between red and black, so your chance is $\frac{18}{38}$.

 What is the chance of the ball landing on an odd number in a roulette wheel? (Note 0 and 00 are neither odd nor even.)

 What is the chance of you betting on a dozen (say 1 to 12) and winning?

18 Which has a higher chance of winning: a dozen or a column (of 12)? Or is the chance the same?

19 What's the chance of betting on a "low" (lower half) in a low-high bet and winning?

Solutions to Problems in Against All Odds: Probability in Gaming

(1) False. It does matter how many numbers there are to choose from in a lottery. Each time another number is added, the number of possibilities increases, and the chance of winning decreases.

(2) False. Buying $500 worth of lottery tickets (100 tickets at $5 each) is a bad idea. It does increase your chance 100 times, but what is 100 times a tiny, tiny number? Still a tiny number.

(3) False. A set of numbers like 35, 36, 37, 38, 39, and 1 has the same chance as any other combination.

(4) True. Before playing the lottery, it's important to check and think about how many numbers there are to pick from; it makes a difference.

(5) The equation for the number of two-card hands is $_{52}C_2 = \dfrac{52!}{2!\,(52-2)!} = \dfrac{52(51)(50!)}{2(50!)} = 1{,}326$.

(6) You need an ace and a ten-point card for the numerator, and 1,326 for the denominator.

First, you need an ace, and there are $_4C_1 = \dfrac{4!}{1!\,(4-1)!} = \dfrac{4*3!}{1*3!} = 4$ ways to get that. Then you need a ten-point card, and there are $_{16}C_1 = \dfrac{16!}{1!\,(16-1)!} = \dfrac{16*15!}{1*15!} = 16$ ways to do that. Putting these together, you get $4*16 = 64$ for the numerator. Dividing by 1,326 gives you 0.048 or 4.8 percent.

(7) You need two ten-point cards, of which there are 16 (four 10s, four Jacks, four Queens, and four Kings). So the equation is $_{16}C_2 = \dfrac{16!}{2!\,(16-2)!} = \dfrac{16(15)(14!)}{2(14!)} = 120$.

(8) Take the number of 20-point hands divided by the total number of two-card hands and you get $^{120}/_{1,326} = 0.090$ or 9.0 percent.

(9) To get a total of 8, you can get 5,3; 3,5; 2, 6; 6, 2; or 4,4. So there are five ways (out of 36), $= \, ^5/_{36}$.

(10) The most difficult rolls are 2 and 12, each of which can happen in only one way (1,1) or (6,6).

(11) The number of ways to get an even total in craps is the same as the number of ways to get 2, 4, 6, 8, 10, or 12. There are 1, 3, 5, 5, 3, and 1 ways to get each, respectively, for a total of 18 ways.

(12) Three of a Kind: 54,912 are possible since you choose the kind (one kind from 13 possible), then take three of the four cards available of that kind, then of the remaining 12 kinds, you choose two (since they can't match), and of the four cards available of each of those two kinds, you take one. The number of ways is $(_{13}C_1 *\,_4C_3)*(_{12}C_2 *\,_4C_1 *\,_4C_1) = 13*4*66*4*4 = 54{,}912$.

(13) Four of a Kind: Of the 13 kinds, you choose one, then of the four cards available, you take all four. Then the remaining card has to be from the other 12 kinds, so you choose that kind, then take one card of the four available: $(_{13}C_1 *\,_4C_4)*(_{12}C_1 *\,_4C_1) = 13*1*12*4 = 624$.

(14) Four of a Kind with face cars only (J, Q, K): There are only three kinds available, not 13, so you pick one of those three kinds, and take all four cards available, then of the remaining non-face card kind, of which there are twelve, you pick one, and take one of the four cards available: $(_3C_1) * (_4C_4) * (_{12}C_1) * (_4C_1) = 3 * 1 * 12 * 4 = 144$.

(15) Four of a Kind with aces only is AAAA____, where the blank is filled with one of the 48 cards that are not an ace, so it's $(_1C_1 * _4C_4) * (_{12}C_1 * _4C_1) = 48$.

(16) The chance is $^{18}\!/_{38}$ since there are 38 numbers total (including 0 and 00) and there are 18 odd numbers: 1, 3, 5, . . ., 35.

(17) The chance is $^{12}\!/_{38}$ since a dozen is 12, and there are 38 numbers total.

(18) A dozen bets on 12 numbers and a column bets on 12 numbers, so their chances are the same: $^{12}\!/_{38}$.

(19) The lower half of the numbers is $^{18}\!/_{38}$ since you bet on 19 to 36, of which there are 18 numbers out of 38 total (including 0 and 00).

3

From A to Binomial: Basic Probability Distributions

Explore discrete random variables and the discrete uniform distribution.

Calculate the mean, the variance, and the standard deviation of X.

Identify the characteristics of the binomial distribution and find probabilities for it.

Explore the bell shape of the normal distribution and find probabilities for the normal and standard normal distributions.

Use the normal distribution to approximate the binomial distribution.

Find probabilities for the sample mean using the Central Limit Theorem.

Explore probability's role in calculating confidence intervals and conducting hypothesis tests.

Chapter **7**

Dealing with Discrete Probability Distributions

In this chapter, you work with discrete random variables, which have either a finite list of possible values or a countably infinite list of possible values. You see how to combine the probabilities and the values to obtain the probability mass function (pmf). You also see how to calculate the mean, the variance, and the standard deviation of X. In addition, you work with the discrete uniform distribution, one of the more common and straightforward discrete distributions. These are all the basic ideas that go along with a random variable, in the discrete case.

Finding the Probability Distribution of a Discrete Random Variable

A *probability* is a function from the sample space, S, to the set of all possible real numbers in the interval [0, 1], denoted by capital letters like X, Y, and so on. For example, roll a die once and let X be the outcome. X can take on the value 1 with probability $P(1) = \frac{1}{6}$. It can also take on the values 2, 3, 4, 5, or 6, all with probability ⅙. If you are flipping a coin two times and counting the number of heads, X can be 0, 1, or 2 with probabilities ¼, ½, and ¼. These are called *probability distributions*. (Note all coins and dice are deemed to be fair throughout this chapter.)

REMEMBER

A *discrete random variable* is a random variable that takes on either a finite number of possible values, like the ones just mentioned, or a countably infinite number of possible values, such as values like 0, 1, 2, 3, and so on. For example, X might be the number of people in line at the bank at any point in time.

The probability distribution of a discrete random variable is a listing of all the possible values of the variable, along with their probabilities. Two conditions must be met for all discrete probability distributions:

>> Each probability must lie between 0 and 1.

>> The total sum of all the probabilities must be 1.

For example, if X is the number of heads that appear in two flips of a coin, the list of possible values (sample space) is S = {0, 1, 2}, and the probability distribution is outlined in the following table:

X	0	1	2
P(x)	P(TT) = ¼	P(HT) + P(TH) = 2/4	P(HH) = ¼

You can see that the table represents a legitimate probability distribution because the two conditions are met.

EXAMPLE

Q. Write out the probability distribution for X = the number of heads in three flips of a fair coin.

A.

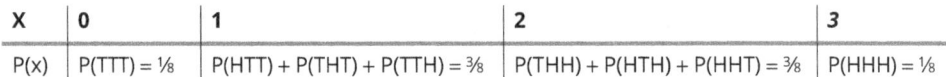

X	0	1	2	3
P(x)	P(TTT) = ⅛	P(HTT) + P(THT) + P(TTH) = ⅜	P(THH) + P(HTH) + P(HHT) = ⅜	P(HHH) = ⅛

Note that all the probabilities are between 0 and 1 and they sum to 1, so it is a legitimate probability distribution.

1. Tell whether the following statement is true or false: "A probability distribution is a list of all possible values of X and their probabilities."

2. Fill in the blank with the best answer: If you roll two die, and X is the sum of the die, there are _____ possible values for X.

3. Tell whether the following statement is true or false: "X values can be negative, but P(x) cannot."

4. Flip a fair coin four times and let X = the number of heads. What are the possible values of X?

Calculating Mean, Variance, and Standard Deviation of a Discrete Random Variable

This section contains formulas to calculate the mean, variance, and standard deviation of a discrete random variable. These values are called parameters and represent the mean, variance, and standard deviation of the entire population of values, and are denoted by Greek letters:

» μ_x represents the mean.

» σ_x^2 represents the variance.

» σ_x represents the standard deviation.

REMEMBER

It's important to note that these values are not the same as the mean, variance, and standard deviation of a data set, which are denoted by $\bar{x}$, s^2, and s, respectively.

To calculate the mean of X when X is a discrete random variable, you use the probability distribution for X. You multiply each value of X by its weight, which is the probability, $p(x)$, and then sum all the values. Values of X that have a higher probability get a higher weight in terms of the mean, and values of X that have a lower probability get a lower weight in terms of the mean.

The formula for the mean of X is $\mu_x = \sum_{\text{all x}} xp(x)$.

For example, consider the following probability distribution for X:

X	1	2	3
P(x)	.5	.3	.4

The mean of X is calculated by $1(0.5) + 2(0.3) + 3(0.4) = 2.3$. That means the mean of X in this case is between 2 and 3, and closer to 2 than 3.

WARNING

A common mistake when calculating the mean of a random variable X is to just average all the possible values of X. This gives all the values the same weight, and doesn't take the $p(x)$ values into account.

To calculate the variance of X when X is a discrete random variable, you use the following formula: $\sigma_x^2 = \sum_{\text{all x}} (x - \mu_x)^2 p(x)$. You take each value of X, subtract the mean, square it to make it positive, then multiply by $p(x)$ to give it weight to indicate how often you see that value of X in the population. In the earlier table, the variance of X is

$$\sigma_x^2 = \sum_{\text{all x}} (x - \mu_x)^2 p(x) = (1 - 2.3)^2(0.5) + (2 - 2.3)^2(0.3) + (3 - 2.3)^2(0.4) = 1.07.$$

The standard deviation of X is the square root of the variance, and is denoted by $\sigma_x = \sqrt{\sum_{\text{all x}} (x - \mu_x)^2 p(x)}$. In this example, the standard deviation is 1.03. That is, the "average" amount of variability from the mean in the population is 1.03.

Q. Suppose that you flip a coin two times and count the number of heads, X. What is the mean of X?

EXAMPLE **A.** First, write down the probability distribution of X:

X	0	1	2
P(x)	¼	½	1/4

The mean of X here is $\mu_x = \sum_{\text{all } x} xp(x) = 0(1/4) + 1(1/2) + 2(1/4) = 1$. That's the overall average number of heads on two flips of a coin.

---

5 What is the variance and standard deviation of the number of heads on two flips of a coin?

6 What is the mean of X, where X is the outcome of a single die?

---

7 What is the variance and standard deviation of X, where X is the outcome of a single die?

8 Given the following probability distribution for X:

x	10
P(x)	1

a. Find the mean of X.

b. Find the variance of X.

c. Find the standard deviation of X.

9 Tell whether the following statement is true or false: "The mean of a discrete random variable X can be negative." If true, give an example; if false, explain why.

10 Tell whether the following statement is true or false: "The variance of a discrete random variable can be negative." If true, give an example; if false, explain why.

11 Tell whether the following statement is true or false: "The standard deviation of a discrete random variable can be negative." If true, give an example; if false, explain why.

12 Given the following probability distribution for X:

x	−1	0	1
P(x)	1/3	1/3	1/3

a. Find the mean of X.

b. Find the variance of X.

c. Find the standard deviation of X.

Exploring the Discrete Uniform Distribution

One of the more common and straightforward discrete distributions is the discrete uniform distribution. You saw an example of it earlier in this chapter: the roll of a single die. Each value of X has the same probability (that's why it's called *uniform*; the values have uniform probability). It's *discrete* because it takes on values from a, a + 1, a + 2, . . ., b, with no probability between the integers of the domain. For example, the discrete uniform distribution can take on values such as 0, 1, 2, 3, 4, 5, each with probability ⅙, with no probability between the values. The graph of the probability mass function (pmf) of this example is shown in Figure 7-1.

FIGURE 7-1:
The graph of a discrete uniform distribution between 0 and 5.

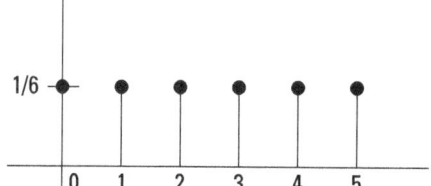

The number of values of X in a discrete uniform distribution is n, where X goes from a to b, and n = b − a + 1 items. For example, if X is 0, 1, 2, 3, 4, 5, we have a = 0, b = 5, and n = b − a + 1 = 5 − 0 + 1 = 6 items, each with probability $\frac{1}{n} = \frac{1}{6}$.

The mean of the discrete uniform distribution is $\frac{(a+b)}{2}$, and the variance is $\frac{(b-a+1)^2-1}{12}$. For our example with values 0, 1, 2, 3, 4, 5, the mean is $\frac{(0+5)}{2} = 2.5$ and the variance is $\frac{(b-a+1)^2-1}{12} = \frac{(5-0+1)^2-1}{12} = \frac{35}{12} = 2.92$. The standard deviation is the square root of the variance, which is $\sqrt{2.92} = 1.71$.

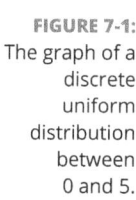

Q. Find the mean, variance, and standard deviation of X, where X is the outcome of one roll of a die.

EXAMPLE

A. Here a = 1, b = 6, n = b − a + 1 = 6 − 1 + 1 = 6, and all probabilities are ⅙. The mean is $\frac{(a+b)}{2} = \frac{(1+6)}{2} = 3.5$. The variance is $\frac{(b-a+1)^2-1}{12} = \frac{(6-1+1)^2-1}{12} = \frac{35}{12} = 2.92$. The standard deviation is the square root of the variance, which is $\sqrt{2.92} = 1.71$.

---

 13   Answer yes or no to the following question: When you flip a coin two times and look at the total number of heads, X, does X have a discrete uniform distribution?

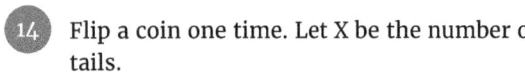 Flip a coin one time. Let X be the number of tails.

    a. What is the mean of X?

    b. What is the variance of X?

    c. What is the standard deviation of X?

15 Answer yes or no to the following question: Suppose that you roll a die twice and take the sum of the outcomes, X. Does X have a uniform distribution?

16 Answer yes or no to the following question: Can the mean of a discrete uniform be negative?

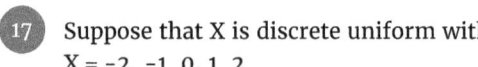 Suppose that X is discrete uniform with X = -2, -1, 0, 1, 2.

    a. What is the mean of X?

    b. What is the variance of X?

Solutions to Problems in Dealing with Discrete Probability Distributions

(1) True. This is the definition of a probability distribution.

(2) The answer is 11: 2, 3, 4, 5, 6, 7, 8, 9, 10, 11, 12.

(3) True. Values of X can be negative, but P(x) probabilities must always lie between 0 and 1.

(4) X can be 0, 1, 2, 3, 4 if you flip a coin four times and count the number of heads.

(5) The variance and standard deviation of the number of heads on two flips of a coin are found using the following equation: $\sigma_x^2 = \sum_{\text{all } x}(x - \mu_x)^2 p(x)$, where the probabilities are ¼, ½, and ¼ for 0, 1, and 2 heads, respectively. The mean number of heads is 1 (taking ¼ * 0 + ½ * 1 + ¼ * 2 = 1), so the variance is $(0-1)^2(¼) + (1-1)^2(½) + (2-1)^2(¼) = ½ = 0.50$. The standard deviation is the square root, which is 0.71.

(6) The probability distribution for X in this case is ⅙ for P(x) for each value of X from 1, 2, . . ., to 6. The mean is 3.5: 1(⅙) + 2(⅙) + . . . + 6 (⅙) = 3.5.

(7) The variance of X in this case is $(1-3.5)^2(⅙) + (2-3.5)^2(⅙) + ... + (6-3.5)^2(⅙) = 2.92$. The standard deviation is the square root of 2.92, which is 1.71.

(8) Given the following probability distribution, there is only one possible value for X, and that is 10. It occurs in the population with a probability of 1, as shown in the following table.

a. The mean of X is 10(1) = 10.

b. The variance of X is 0: $(10-10)^2(1) = 0$.

c. The standard deviation is the square root of the variance, which is also 0.

(9) The mean of a discrete random variable X can be negative. An example is illustrated in the following table. The mean is -1 * ½ + -2(½) = -1.5.

x	-1	-2
P(x)	½	½

(10) The variance of a discrete random variable can never be negative because you square all of the differences to make them nonnegative in the formula.

(11) The standard deviation of a discrete random variable can never be negative because it's a (principal) square root.

(12) Given the following probability distribution for X:

x	-1	0	1
P(x)	⅓	⅓	⅓

a. The mean of X is $-1 * \frac{1}{3} + 0 * \frac{1}{3} + 1 * \frac{1}{3} = 0$.

b. The variance of X is $(-1-0)^2(\frac{1}{3}) + (0-0)^2(\frac{1}{3}) + (1-0)^2(\frac{1}{3}) = \frac{2}{3}$.

c. The standard deviation of X is the square root of the variance, the square root of ⅔, which is 0.816.

(13) No. The probabilities are not all the same for the X values. $P(0) = (\frac{1}{4})$, and $P(2) = (\frac{1}{4})$, but $P(1) = (\frac{1}{2})$. This is not a uniform distribution.

(14) $X = 0, 1; n = (1 - 0 + 1) = 2$.

a. The mean is $\frac{(a+b)}{2} = \frac{(0+1)}{2} = \frac{1}{2}$.

b. The variance is $\frac{(b-a+1)^2 - 1}{12} = \frac{(1-0+1)^2 - 1}{12} = \frac{3}{12} = \frac{1}{4}$.

c. The standard deviation is the square root of ¼, which is 0.5.

(15) No. The outcomes in the middle of the distribution have higher probabilities, and the outcomes on the low or high end of the distribution have lower probabilities. It's not uniform.

(16) Yes. If at least some of the values of X are negative.

(17) X is discrete uniform with X = -2, -1, 0, 1, 2.

a. The mean is $\frac{(a+b)}{2} = \frac{-2+2}{2} = 0$.

b. The variance is $\frac{(b-a+1)^2 - 1}{12} = \frac{(2--2+1)^2 - 1}{12} = \frac{24}{12} = 2$.

IN THIS CHAPTER

» Identifying a binomial distribution

» Finding probabilities for the
binomial distribution

» Calculating the mean and variance
of the binomial distribution

Chapter **8**

Juggling Success and Failure with the Binomial Distribution

The binomial distribution is used when the data fall into one of two groups: yes (success) and no (failure). The idea is to count the number of yeses that result from n "trials," or n observations of a random phenomenon. Like flipping a coin, for example. You can get heads or tails. If you want to count the number of heads in ten flips, you call a head a "yes" or a "success" and a tail a "no" or a "failure."

You can find probabilities of getting certain results, such as the probability of getting eight heads out of ten flips (which you'd guess is pretty small), or the probability of getting five heads out of ten flips (which has a higher chance of happening).

In this chapter, you first look at the characteristics of the binomial distribution and find probabilities for it, and then practice calculating the mean, variance, and standard deviation of it.

Identifying the Characteristics of the Binomial Distribution

A binomial distribution has the following four characteristics:

>> There are n (a fixed number of) observations of a random phenomenon, also known as *trials*.

>> Each trial has one of exactly two outcomes: yes (success) or no (failure).

>> The n trials are independent, meaning they don't affect each other.

>> The chance of getting a success is p, and stays the same for each trial.

The random variable X stands for the number of successes in n trials. For the coin flipping example, X = the number of heads (successes) you get when you flip a fair coin ten times. There are n = 10 trials; each trial has one of two possible outcomes: heads or tails. The ten trials (flips) are independent, and the probability of a success (head) is p = ½ and stays the same for each flip (assume the coin is fair). Therefore, X has a binomial distribution.

EXAMPLE

Q. You flip a fair coin until you get three heads. Let X be the number of flips you need. Does X have a binomial distribution?

A. No. The value of n is unknown, so it's not fixed.

---

1 Answer yes or no to the following question: You different set of horses race a series of n = 10 random races, and X = the number of times the horse you pick wins the race. Does X have a binomial distribution?

2 Answer yes or no to the following question: You roll a six-sided fair die 20 times and record the outcome each time. X = the number of times a 6 comes up. Does X have a binomial distribution?

3 Answer yes or no to the following question: You randomly pull 100 M&M candies out of a large bag and record X = the color of the M&M selected. Does X have a binomial distribution?

4 Answer yes or no to the following question: You randomly pull 100 M&M candies out of a large container at the factory and record the color of each M&M. You count X = number of red M&Ms you pull out. Does X have a binomial distribution?

5 Answer yes or no to the following question: You randomly pull M&Ms out of a large bag until you get ten red ones. X is the number of M&Ms pulled out. Does X have a binomial distribution?

6 Answer yes or no to the following question: You randomly select ten people and record their heights. Then you count the number of people in your sample who are under 5.5 feet tall and call that number X. Does X have a binomial distribution? (Note that it is estimated that 25 percent of people are under 5.5 feet tall.)

Answer yes or no to the following question: You randomly select a person and measure their height. Then you measure the heights of all their friends. You count the number of people in the group who are over 5.5 feet tall and call that number X. Does X have a binomial distribution? (*Note:* Suppose that 25 percent of people are under 5.5 feet tall.)

 Answer yes or no to the following question: You have a population of 100 people that is 50 percent male and 50 percent female. You select one person at random and record whether or not that person is female. You take this person out of the population and select another person at random from the remaining group. You keep doing this 20 times. Let X = number of females. Does X have a binomial distribution?

Finding Probabilities for the Binomial Distribution

To find the probability of getting X successes among n fixed trials, each having the same probability, p, of getting a success, you use the following formula: $p(x) = \binom{n}{x} p^x (1-p)^{n-x}$, where x is 0, 1, 2, . . ., n.

The first part of the formula, $\binom{n}{x}$ is telling us $\binom{n}{x} = \frac{n!}{x!\,(n-x)!}$, where $n! = n(n-1)(n-2)\ldots$ (3)(2)(1). It's the number of ways to count the ways to place the successes (yeses) and failures (nos). For example, $3! = 3(2)(1) = 6$, and $2! = 2(1) = 2$; $1! = 1$ and by convention, $0! = 1$ also. If you need a single probability for the binomial, this formula is a good way to go.

The binomial table (Table A-1) in the appendix shows cumulative probabilities for the binomial; in other words, it sums them up for you. So if you want $P(X \le 3)$, the table gives you $P(X = 0) + P(X = 1) + P(X = 2) + P(X = 3)$. For example, if p = 0.5 and n = 10, $P(X \le 3) = 0.172$. To find this in the table, go to the section of the table for n = 10, then look at the column where p = 0.5, and the row where x = 3.

If you want to find less-than, greater-than, or greater-than-or-equal-to probabilities, you can manipulate the values in the table to find the probabilities you want. For example, $P(X \geq 4) = 1 - P(X \leq 3) = 1 - 0.172 = 0.828$.

Q. Using the probability formula, find $P(X = 3)$ when $n = 10$ and $p = 0.2$.

A. $P(X = 3) = \binom{10}{3} \cdot 2^3 (1 - .2)^{10-3} = \dfrac{10 * 9 * 8 * 7!}{3!7!}(0.008)(0.209715) = 120 * 0.001678 = 0.2013$

EXAMPLE

---

9 Flip a fair coin ($p = 0.5$) ten times ($n = 10$). Let X = the number of heads. What's the probability that X is equal to 4?

10 Flip a fair coin ($p = 0.5$) ten times ($n = 10$). Let X = the number of heads. What's the probability that X is equal to 6?

---

11 Flip a fair coin ($p = 0.5$) ten times ($n = 10$). Let X = the number of heads. What's the probability that X is greater than or equal to 4?

12 Check to see how many cars in a random sample of 8 roll through a stop sign at an intersection, rather than come to a complete stop. Previous research says $p = 0.30$. What's the chance that X is at least 2?

13 Roll a fair die ten times. What's the chance of getting at most two 6s?

14 Roll a fair die ten times. What's the chance of getting at least three 6s?

15 Suppose that 90 percent of college students graduate from the same school at which they started. Take a random sample of ten college students. What's the chance of getting fewer than or equal to eight of the students who will graduate from the same school at which they started?

16 Suppose that 20 percent of pets are named Buddy. You take a random sample of 20 pets and record their names. What's the chance that fewer than three of them are named Buddy?

Finding the Mean, Variance, and Standard Deviation of the Binomial Distribution

The mean of the binomial distribution has a formula given by np. It's the proportion of yeses you can expect (p) times the number of trials (n).

The variance of the binomial is $np(1-p)$. You can see it fluctuates between the probability of success (p) and the probability of failure $(1-p)$, n times.

The standard deviation is the square root of the variance. Its formula is $\sqrt{np(1-p)}$.

EXAMPLE

Q. Find the mean of X (binomial) when n = 10 and p = 0.2.

A. The mean of X is np = 10(0.20) = 2.

---

17 Suppose that 95 percent of students in sports majors graduate. You take a random sample of 20 students in a sports major. What is the mean, variance, and standard deviation of the number (X) of students who will graduate?

18 Suppose that 10 percent of gardeners grow sunflowers in their gardens. You take a random sample of ten gardens. What is the mean, variance, and standard deviation of the number of gardens that have sunflowers?

19 Research tells us that 20 percent of people wear hats most of the time. You take a random sample of 30 people. What's the mean, variance, and standard deviation of the number of people who wear hats most of the time?

20 You hear that 25 percent of cars are blue. You take a sample of 15 cars, and X = the number of cars that are blue. Find the mean, variance, and standard deviation of X.

21 Find the mean, variance, and standard deviation of the number of 2s you get when you roll a die 100 times.

22 Find the mean, variance, and standard deviation of the number of correct answers you get if you guess randomly on 20 multiple-choice questions, where each question has four possible answers.

Solutions to Problems in Juggling Success and Failure with the Binomial Distribution

(1) No, X does not have a binomial distribution. You have n = 10 trials that are independent because they are randomly chosen. The two outcomes you are interested in are "the horse you picked wins" and "the horse you picked loses." However, the value of p = the probability the horse you picked wins differs with each horse and with each race. So X does not have a binomial distribution.

(2) Yes, X has a binomial distribution. You have n = 20 observations (rolls) that are independent. Success = "a six comes up," and failure = "a six doesn't come up." The value of p = ⅙ and is the same for each roll. Therefore, X has a binomial distribution.

(3) No, X does not have a binomial distribution. You have n = 100 M&Ms (trials). But you record X = the color of the M&M. X doesn't have only two outcomes; it has several outcomes, one for each color. So X does not have a binomial distribution.

(4) Yes, X has a binomial distribution. You have n = 100 M&Ms (trials). You count X = the number of red M&Ms, so the two outcomes are "red" and "non-red." The probability, p, of a red M&M is 13 percent, according to Mars, the company that makes M&Ms. The value of p stays the same because the container is extremely large. So X does have a binomial distribution.

(5) No, X does not have a binomial distribution. The keyword is "until." You don't know how many fixed trials you have because you keep sampling *until* you have ten red ones. So X does not have a binomial distribution.

(6) Yes, X has a binomial distribution. While height itself is not binomial, you are counting the number of people in your sample who are under 5.5 feet tall, so that's a yes or no characteristic. You have n = 10 and p = 0.25. So X does have a binomial distribution.

(7) No, X does not have a binomial distribution. Because you are measuring only one randomly chosen person and then sampling all of that person's friends, you do not have a random sample. So X does not have independence, and you do not have a binomial distribution.

(8) No, X does not have a binomial distribution. You have a population of only 100 people. Suppose that 50 percent of the population, or 50 people, are female. If you draw a female, the next value of p is 49/99, which is not 50 percent. The value of p keeps changing every time you select someone, whether they are female or male. So X does not have a binomial distribution.

(9) $P(X = 4)$ where n is binomial with n = 10 and p = 0.5 is $\binom{10}{4}0.5^4(1-0.50)^6 = \frac{10!}{4!\,(6!)}0.00098 =$ $\frac{10(9)(8)(7)(6!)}{4(3)(2)(1)(6!)}0.00098 = \frac{5040}{24}0.00098 = 0.2058$.

(10) $P(X = 6)$ where n is binomial with n = 10 and p = 0.5 is $\binom{10}{6}0.5^6(1-0.50)^4 = \frac{10!}{6!\,(4!)}0.00098 =$ $\frac{10(9)(8)(7)(6!)}{4(3)(2)(1)(6!)}0.00098 = \frac{5040}{24}0.00098 = 0.2058$.

TIP

Note that the solutions for questions 9 and 10 are the same and give the same answer. That's because p = 0.50, which makes the binomial symmetric. It will give the same answers for 4 and 6, for 3 and 7, for 2 and 8, for 1 and 9, and for 0 and 10.

(11) You have n = 10, p = 0.5, and you want $P(X \geq 4)$. The binomial table in the appendix gives probabilities in the form of $P(X \leq x)$, so you'd rewrite your probability to get $P(X \geq 4) = 1 - P(X \leq 3) = 1 - 0.172 = 0.828$.

(12) For this problem, n = 8 and p = 0.30. The chance that X is at least two is written as $P(X \geq 2) = 1 - P(X \leq 1) = 1 - 0.255 = 0.745$.

(13) Here n = 10, and you want $P(X \leq 2)$, but you have to determine p. A success is getting a 6. The chance of that is $\frac{1}{6}$, so that is p. You can't use the binomial table in the appendix for this because p = 0.167 is not one of the values on the table. Instead, use the binomial probability formula and get $P(X \leq 2) = P(X = 0) + P(X = 1) + P(X = 2)$. You get $\binom{10}{0} 1/6^{0}(5/6)^{10} +$ $\binom{10}{1} 1/6^{1}(5/6)^{9} + \binom{10}{2} 1/6^{2}(5/6)^{8} = 1(0.1615) + 10(0.1667)(0.1938) + 45(0.0278)(0.2326) = 0.7755$.

(14) Here n = 10 and you want $P(X \geq 3)$. Note that this is the opposite, or complement, of the previous problem. So you can take 1 minus the answer to the previous problem. This gives you $1 - 0.7755 = 0.2245$.

(15) You have n = 10, p = 0.90, and $P(X \leq 8)$. According to the binomial table in the appendix, the answer is 0.264.

(16) For this problem, n = 20 and p = 0.20. You want $P(X < 3)$. This is rewritten as $P(X \leq 2) = 0.206$ according to the binomial table.

(17) In this scenario, n = 20 and p = 0.95. That's all you need to calculate the mean, variance, and standard deviation. The mean is np = 20(0.95) = 19 sports majors. The variance is np(1 – p) = 20(0.95)(0.05) = 0.95 (no units), and the standard deviation is the square root of 0.95, which is 0.97 for sports majors.

(18) Here, n = 10 and p = 0.10. The mean is np = 10(0.10) = 1 garden. The variance is np(1 – p) = 10(0.10)(0.90) = 0.90 (no units), and the standard deviation is the square root of 0.90, which is 0.95 gardens.

(19) In this problem, n = 30 and p = 0.20. The mean is np = 30(0.20) = 6 people. The variance is np(1 – p) = 30(0.20)(0.80) = 4.80 (no units), and the standard deviation is the square root of 4.80, which is 2.19 people.

(20) Here, n = 15 and p = 0.25. The mean is np = 15(0.25) = 3.75 cars. The variance is np(1 – p) = 15(0.25)(0.75) = 2.81 (no units), and the standard deviation is the square root of 2.81, which is 1.68 cars.

(21) You have n = 100, and p is the chance of getting a 2 when you roll a die, which is $\frac{1}{6}$. The mean is np = 100 ($\frac{1}{6}$) = 16.67 twos. The variance is np(1–p) = 100($\frac{1}{6}$)($\frac{5}{6}$) = 13.89 (no units), and the standard deviation is the square root of 13.89, which is 3.73 twos.

(22) If you guess randomly on a multiple-choice question that has four answers, the chance of getting the correct answer is $\frac{1}{4}$ or 0.25. There are n = 20 questions. The mean is np = 20(0.25) = 5 questions right. The variance is np(1 – p) = 20(0.25)(0.75) = 3.75 (no units), and the standard deviation is the square root of 3.75, which is 1.94 correct answers.

Chapter **9**

Normalizing the Normal Distribution

The normal distribution is a widely used probability model for many random phenomena whose results follow the well-known bell-shaped pattern. Its uses go from modeling real-world phenomena to quality-control scenarios. To understand how it works is very important for statistics.

In this chapter, you explore the normal distribution's bell-shaped pattern and how to use it to answer questions about probabilities for the normal distribution. As part of the process, you examine and use the standard normal distribution. You also practice using the normal distribution to find percentiles.

REMEMBER

A percentile is a value of X such that a given percentage of values lie below it. For example, if your results are in the 99th percentile on a national achievement test, 99 percent of the people taking the test scored lower than you.

Charting the Basics of the Normal Distribution

The *normal distribution* is known as the "bell-shaped" distribution, or the "bell-shaped curve." It is symmetric and looks like a bell. The mean is designated by μ and is the middle value of the distribution. The standard deviation is designated by σ and there are three standard deviations of values on either side of the mean (shown by tick marks) that account for most of the values. About 68 percent of the values are within one standard deviation of the mean, about 95 percent of the values are within two standard deviations of the mean, and about 99.7 percent of the values are within three standard deviations of the mean. On each end of the normal distribution is a tail that goes down on each side.

For example, if X has a normal distribution with a mean of 75 and a standard deviation of 5, X would look like the bell curve shown in Figure 9-1. Every normal distribution has its own mean and standard deviation.

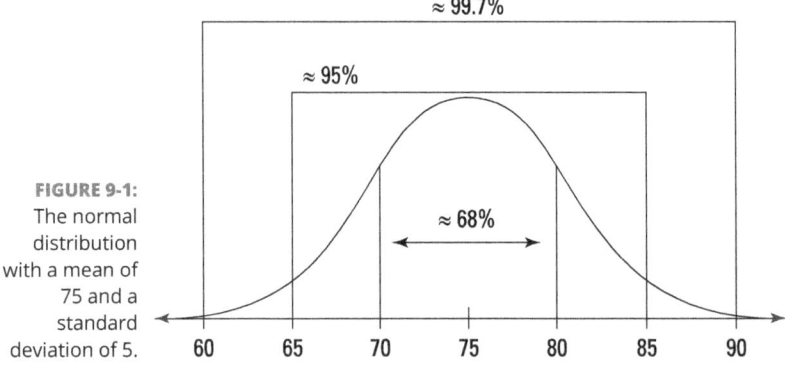

FIGURE 9-1:
The normal distribution with a mean of 75 and a standard deviation of 5.

EXAMPLE

Q. What is the standard deviation of the following normal distribution? (Note, the tick marks indicate distances equal to the standard deviation.)

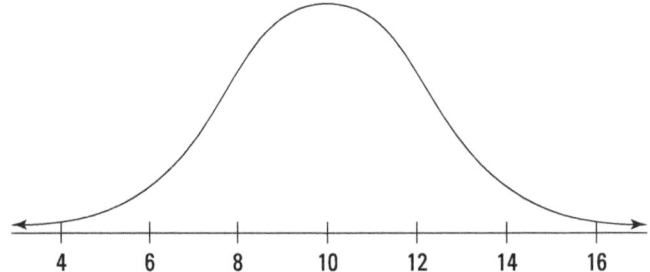

A. The standard deviation is 2. Three tick marks appear on either side of the mean that increase and decrease by 2s.

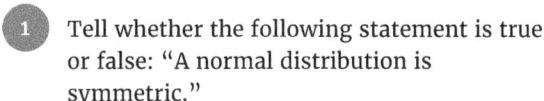 Tell whether the following statement is true or false: "A normal distribution is symmetric."

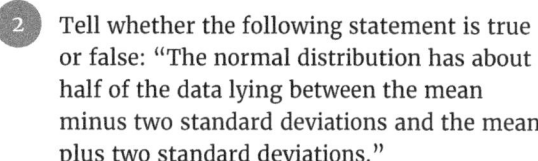 Tell whether the following statement is true or false: "The normal distribution has about half of the data lying between the mean minus two standard deviations and the mean plus two standard deviations."

3 Suppose that exam scores have a normal distribution with a mean of 80 and a standard deviation of 5.

   a. Draw a picture of the normal distribution and label the three tick marks on either side, indicating the standard deviations.

   b. About 68 percent of the values lie between what two numbers on this distribution?

   c. About 95 percent of the values lie between what two numbers on this distribution?

   d. About 99.7 percent of the values lie between what two numbers on this distribution?

Understanding the Standard Normal (Z) Distribution

One special normal distribution is the *standard normal distribution*. It has a mean of 0 and a standard deviation of 1. It's also called the *Z-distribution*. It's special because its values represent the number of standard deviations you are below or above the mean. For example, if Z = +2, you are two standard deviations above the mean because each standard deviation is worth 1. If Z = -2, you are two standard deviations below the mean. If Z = 0, you are right at the mean. 99.7 percent of the values of the Z-distribution lie between -3 and +3 (see Figure 9-2).

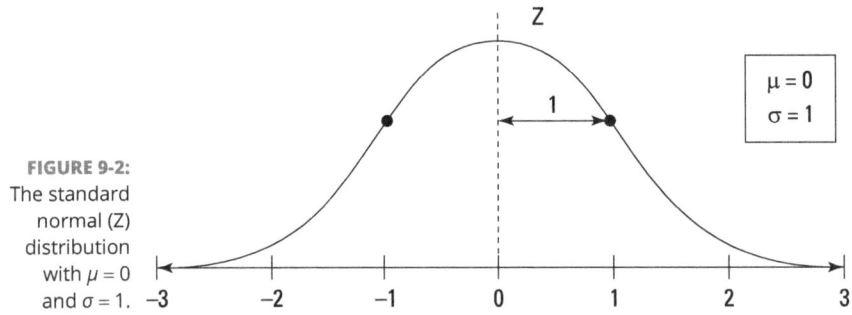

FIGURE 9-2:
The standard normal (Z) distribution with $\mu = 0$ and $\sigma = 1$.

Values on the Z-distribution are called Z-values, Z-scores, or standard scores. The Z-table (Table A-2 in the appendix) is a table of probabilities for all Z-scores, from about -3 to +3. Because calculating the probabilities for the Z-distribution — which is continuous — takes a lot of math, a Z-table of probabilities was created a long time ago that we still use today. The first several rows of the Z-table are shown in Figure 9-3.

The distribution at the top of Figure 9-3 shows that the table gives you the probability of being less than the Z-value you are looking up. For example, if you want the probability that Z is less than -1, you first convert -1 into a number with two digits after the decimal point, that is, -1.00. The leading digit and the first digit after the decimal point are -1.0, and that number tells you which row of the Z-table you are in. The last digit tells you which column you are in. In this example, the last digit is 0, so you are in row -1.0 and column 0. Intersect that row and column, and you get a probability of 0.1587.

The Z-table values are found in Table A-2 in the appendix.

REMEMBER If you want a greater-than probability, you take one minus the table value. For example, P(Z > -1.00) is 1 – P(Z < -1.00), which is 1 – 0.1587, or 0.8413. To find the probability between two values, you look up each value on the Z-table and subtract the smaller result from the larger result. For example, to get P(-2.00 < Z < -1.00), you find P(Z < -1.00) = 0.1587, and P(Z < -2.00) = 0.0228. Subtract 0.1587 – 0.0228 and you get 0.1359.

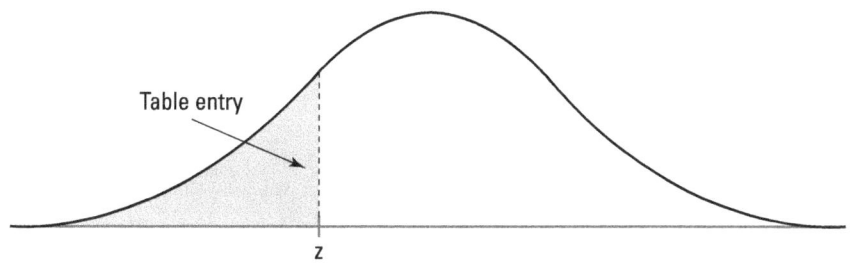

Table entry

The table entry is the area under the standard normal curve to the left of (less than) the z-score.

z	0	0.02	0.03	0.04	0.05	0.06	0.07	0.08	0.09
−3.4	0.0003	0.0003	0.0003	0.0003	0.0003	0.0003	0.0003	0.0003	0.0002
−3.3	0.0005	0.0005	0.0004	0.0004	0.0004	0.0004	0.0004	0.0004	0.0003
−3.2	0.0007	0.0006	0.0006	0.0006	0.0006	0.0006	0.0005	0.0005	0.0005
−3.1	0.0010	0.0009	0.0009	0.0008	0.0008	0.0008	0.0008	0.0007	0.0007
−3.0	0.0013	0.0013	0.0012	0.0012	0.0011	0.0011	0.0011	0.0010	0.0010
−2.9	0.0019	0.0018	0.0017	0.0016	0.0016	0.0015	0.0015	0.0014	0.0014
−2.8	0.0026	0.0024	0.0023	0.0023	0.0022	0.0021	0.0021	0.0020	0.0019
−2.7	0.0035	0.0033	0.0032	0.0031	0.0030	0.0029	0.0028	0.0027	0.0026
−2.6	0.0047	0.0044	0.0043	0.0041	0.0040	0.0039	0.0038	0.0037	0.0036
−2.5	0.0062	0.0059	0.0057	0.0055	0.0054	0.0052	0.0051	0.0049	0.0048
−2.4	0.0082	0.0078	0.0075	0.0073	0.0071	0.0069	0.0068	0.0066	0.0064
−2.3	0.0107	0.0102	0.0099	0.0096	0.0094	0.0091	0.0089	0.0087	0.0084
−2.2	0.0139	0.0132	0.0129	0.0125	0.0122	0.0119	0.0116	0.0113	0.0110
−2.1	0.0179	0.0170	0.0166	0.0162	0.0158	0.0154	0.0150	0.0146	0.0143
−2.0	0.0228	0.0217	0.0212	0.0207	0.0202	0.0197	0.0192	0.0188	0.0183

FIGURE 9-3: The first several rows of the Z-table.

EXAMPLE

Q. Find P(Z < −1.34).

A. Find row −1.3 and column 0.04 on the Z-table and intersect them to get 0.0901.

 If you took an exam and your Z-score was 2.5, what does this mean?

 Find P(−0.01 < Z < −1.5).

6 Find P(Z > -2.22).

7 Find P(Z = 0.00).

8 Find P(Z ≤ 0.00).

9 Suppose that your exam score transformed to a Z-score of 0.00. What does that mean in terms of your exam score?

Finding Probabilities for a Normal Distribution

Finding probabilities for any normal distribution involves math that is way beyond the scope of this book. Still, you can use the Z-table to find probabilities for any normal distribution by changing (or transforming) X to Z, and finding the corresponding probability on the Z-table. After all, you can't have a table for every single normal distribution out there, given that there is an infinite number of them!

To transform X to Z, let's look at an example. Suppose that X has a normal distribution with a mean of 10 and a standard deviation of 2. This X distribution looks like Figure 9-4.

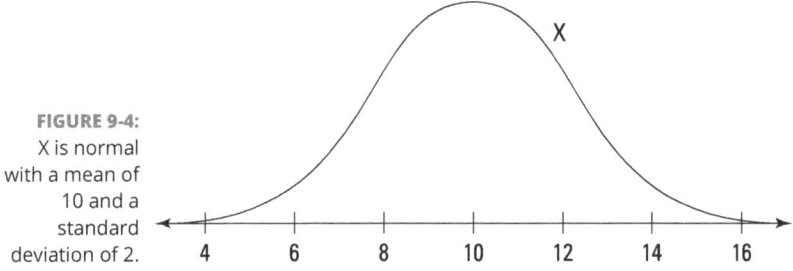

FIGURE 9-4:
X is normal with a mean of 10 and a standard deviation of 2.

In two steps, you can transform this normal distribution to a standard normal (Z) distribution:

1. Subtract the mean, μ, from every value of X (see Figure 9-5).

2. Divide each value of X by the standard deviation of $\sigma = 2$ (see Figure 9-6).

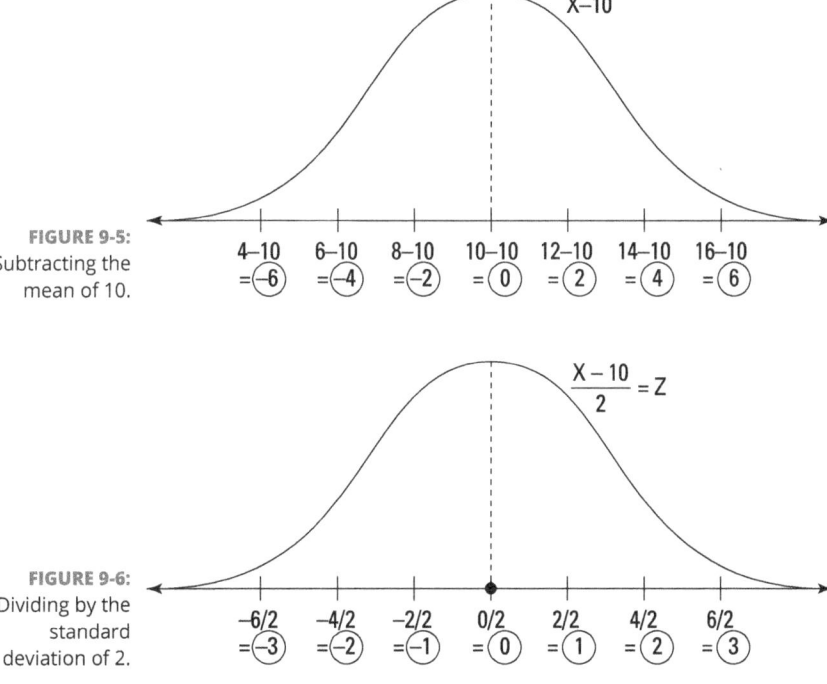

FIGURE 9-5:
Subtracting the mean of 10.

FIGURE 9-6:
Dividing by the standard deviation of 2.

Now you can see that X has been transformed to Z using the formula $Z = \frac{X - \mu}{\sigma}$. The probabilities will be the same once the transformation is finished, and you can look up the transformed value on the Z-table to find the probability. From there, either take the probability if you have a less-than or less-than-or-equal-to scenario, take one minus the probability if you have greater-than or greater-than-or-equal-to scenario, or if there are two values, transform both using Z and subtract the probabilities that you get.

Staying with our example, suppose that you want P(X < 8), where X is normal with a mean of 10 and a standard deviation of 2. This is the transformation in action: $P(X < 8) = P\left(\frac{X - \mu}{\sigma} < \frac{8 - 10}{2}\right) = P(Z < -1) = 0.1587$.

Q. Suppose that X = exam scores that have a normal distribution with a mean of 70 and a standard deviation of 5. Find P(X > 80).

EXAMPLE **A.** $P(X > 80) = P\left(Z > \frac{80 - 70}{5}\right) = P(Z > 2) = 1 - P(Z < 2) = 1 - 0.9772 - 0.0228$.

---

10 Suppose that light bulb lifetimes have a normal distribution with a mean of 100 hours and a standard deviation of 10 hours. Find the probability that a randomly chosen lightbulb lasts more than 125 hours.

11 Bruce knows that his monthly charges on his credit card have a normal distribution with a mean of $200 and a standard deviation of $50. What's the chance that in a randomly chosen month, he charges at most $275?

The passing yards for Flanders' football team in a given game have a normal distribution with a mean of 200 and a standard deviation of 25 yards. What's the chance that in a randomly chosen game, Flanders' team makes between 150 and 250 yards?

13 The filling amounts for frozen yogurt cups have a normal distribution with a mean of 8 ounces and a standard deviation of 0.50 ounces. How often should we get a cup filled with 9 ounces or more?

Handling Percentiles

The kth percentile for X is a value of X, where k percent of the other values lie below it. For example, if you scored in the 70th percentile on an exam, you didn't score a 70 percent, but rather, 70 percent of the other test-takers scored less than you did. A percentile is a value of X (see Figure 9-7).

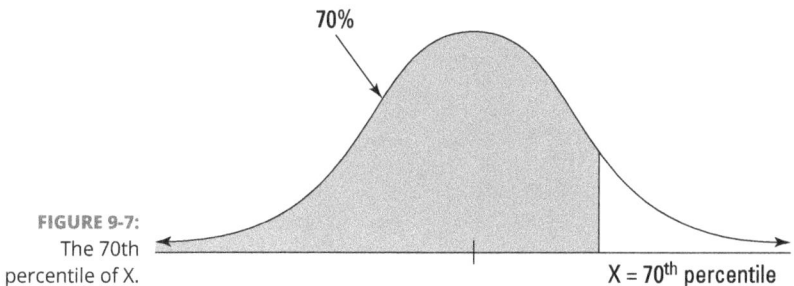

FIGURE 9-7: The 70th percentile of X.

To find the kth percentile for X when X has a normal distribution, you do the following:

1. **Change the value of k to a decimal and look it up in the body of the Z-table.**

2. **Find out which row and column k is in on the Z-table and form the Z-value.**

   This is the kth percentile of the Z-distribution.

3. **Change the value of Z to X using the formula, $Z = \dfrac{X - \mu}{\sigma}$ and solve for X.**

   Note that you can rearrange the Z formula to get $X = Z\sigma + \mu$, which is already solved for X.

   For example, if X has a normal distribution with a mean of 70 and a standard deviation of 5, and you want to find the 90th percentile for X, you look in the table for the closest value to 0.9000, which is 0.8997, and see that it is in the 1.2 row and the 0.08 column. That makes Z = 1.28 the 90th percentile for Z. Now change to X by using the X formula, X = 1.28(5) + 70 = 76.4. This is the 90th percentile for X.

Q. **Suppose that X is normal with a mean of 70 and a standard deviation of 5. What is the tenth percentile of X?**

A. You look up the number closest to 0.1000 in the Z-table and you get 0.1003, which is in row −1.2 and column 0.08, so Z = −1.28 is the 90th percentile for Z. Now change to X by taking X = −1.28(5) + 70 = 63.6 (see Figure 9-8).

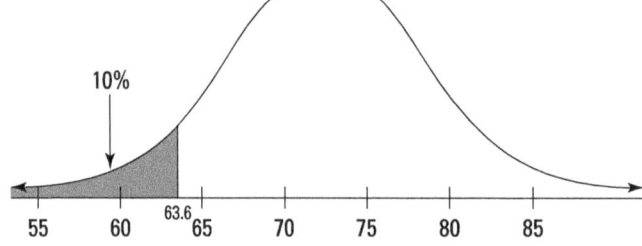

FIGURE 9-8: The tenth percentile of X, where X has a mean of 70 and a standard deviation of 5.

---

14 What is the 50th percentile of X, where X has a normal distribution with a mean of 10 and a standard deviation of 2?

15 What is the 90th percentile of X, where X has a normal distribution with a mean of 10 and a standard deviation of 2?

16 What is the 25th percentile of X, where X has a normal distribution with a mean of 10 and a standard deviation of 2?

17 Suppose that exam scores have a mean of 80 and a standard deviation of 5. Your score is at the 80th percentile. Did you get an 80 on the exam?

Solutions to Problems in Normalizing the Normal Distribution

1. True. The normal distribution is symmetric; it looks the same on both sides when you cut it down the middle.

2. False. It has about 95 percent of the data within that range, much more than half.

3. When exam scores have a normal distribution and a mean of 80 and a standard deviation of 5:

   a. The normal distribution looks like this:

   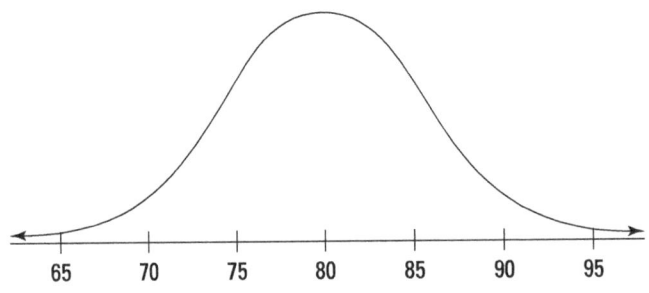

   b. About 68 percent of the scores lie between 75 and 85.

   c. About 95 percent of the scores lie between 70 and 90.

   d. About 99.7 percent of the scores lie between 65 and 95.

4. If you took an exam and your Z-score was 2.5, it means your actual score was 2.5 standard deviations above the mean.

5. To find P(-0.01 < Z < -1.5), you look up P(Z < -0.01) and P(Z < -1.50) on the Z-table and subtract the results: 0.4960 – 0.0668 = 0.4292.

6. To find P(Z > -2.22), you take 1 – P(Z < -2.22). Then look up P(Z < -2.22) on the Z-table to get 0.0132. So, the final answer is 1 – 0.0132 = 0.9868.

7. P(Z = 0.00) = 0 because Z is a continuous random variable and there is no probability at a single point.

8. $P(Z \le 0.00) = 0.5000$.

9. If your exam score had a standard score (Z-score) of 0.00, you are right exactly at the mean itself.

10. You want $P(X > 125) = P\left(Z > \frac{125-100}{10}\right) = P(Z > 2.50) = 1 - 0.9938 = 0.0062$.

11. You want $P(X \le 275) = P\left(Z \le \frac{275-200}{50}\right) = P(Z \le 1.50) = 0.9332$.

12. You are looking for $P(150 < X < 250) = P\left(\frac{150-200}{25} < X < \frac{250-200}{25}\right) = P(-2 < Z < 2) =$
    $P(Z < 2) - P(Z < -2) = 0.9772 - 0.0228 = 0.9544$.

(13) The filling amounts for frozen yogurt cups have a normal distribution with a mean of 8 ounces and a standard deviation of 0.50 ounces. You want $P(X \geq 9) = P\left(Z \geq \frac{9-8}{0.50}\right) = P(Z \geq 2.00) = 1 - 0.9772 = 0.0228$.

(14) The 50th percentile of X is the median, which is also the mean, since the normal distribution is symmetric, so the answer is 10.

(15) To find the 90th percentile of X, you first find the 90th percentile of Z. Look in the body of the Z-table for the number closest to 0.9000, which is 0.8997, and you find it is in row 1.2 and column 0.08, so Z = 1.28. Now change to X with the formula $Z = \frac{X-\mu}{\sigma} \rightarrow 1.28 = \frac{X-10}{2} \rightarrow X = 12.56$.

(16) To find the 25th percentile of X, you first find the 25th percentile of Z. Look in the body of the Z-table for the number closest to 0.2500, which is 0.0.2514, and you find it is in row -0.6 and column 0.07, so Z = -.67. Now change to X with the formula, $Z = \frac{X-\mu}{\sigma} \rightarrow -0.67 = \frac{X-10}{2} \rightarrow X = 8.66$.

(17) No; 80 is the percentile you want, not the score you got. You want the 80th percentile of X. First, find the 80th percentile of Z, which is 0.84 (which is associated with 0.7995 in the body of the table). Then change this to X: $Z = \frac{X-\mu}{\sigma} \rightarrow 0.84 = \frac{X-80}{5} \rightarrow X = 80 + 0.84(5) = 84.2$.

Chapter **10**

Approximating a Binomial with a Normal Distribution

The binomial distribution is popular because a lot of data comes from binomial (yes or no) scenarios. However, calculations with the binomial distribution get unwieldy when n reaches a certain point, and the table values run out, but you can do an approximation. The approximation you use turns out to be the normal approximation if the conditions for np and n(1 – p) hold. This works when n is sufficiently large. The normal approximation involves taking the binomial mean and standard deviation and putting them into the Z formula where $Z = \frac{X - mean}{sd}$ and the mean and standard deviation (sd) come from the binomial. Recall from Chapter 8 that the mean of the binomial distribution is np, and the standard deviation is the square root of np(1 – p).

TIP

While many professors still use the normal approximation to the binomial, which is included here because they don't allow use of the Internet on exams, it makes sense to move to the approximation for values outside of the table for this book. But in general, it's not difficult to get exact binomial probabilities with higher values. There are apps or websites that can do this, such as https://stattrek.com/online-calculator/binomial.

In this chapter, you practice using the normal distribution to approximate the binomial distribution under certain conditions. The results get better as n increases, and the normal distribution starts to take over.

TIP

For a refresher, flip to Chapter 8 for practice with the binomial distribution and Chapter 9 for details on the normal distribution.

Identifying When You Need the Normal Approximation

If you have a binomial problem, there are three ways to solve it, depending on the sample size:

>> You can use the formula for finding probabilities (see Chapter 8), which always works, but can be tedious and time-consuming.

>> You can use the binomial table (Table A-1 in the appendix) for n up through 25.

>> If the sample size is too big for the binomial table or formula, you can check to see whether you can use the normal approximation.

The conditions to use the normal approximation for the binomial are $np \geq 10$ and $n(1-p) \geq 10$. The values of n and p are given in the problem. These conditions basically make sure n is large enough that you will have enough expected yeses (np) and enough expected nos (n(1 – p)) and to balance values out to the middle like the normal distribution does.

TIP

Notice that if p is small, the first condition will be harder to meet than the second condition. If p is small, then (1 – p) will be large. And if p is large, the second condition will be harder to meet because p will be large and (1 – p) will be small. That's why you need both conditions. If $p = \frac{1}{2}$, both conditions give the same results.

EXAMPLE

Q. Suppose that Naomi has done research and finds that 90 percent of college students have worked at a restaurant. She takes a random sample of 200 students. She finds that more than 185 of them have worked at a restaurant. Can Naomi use the normal approximation to the binomial distribution here?

A. $np = 200(0.90) = 180 \geq 10$ and $n(1-p) = 200(1-0.90) = 200(0.10) = 20 \geq 10$. So yes, Naomi can use the normal approximation.

1 Suppose that 70 percent of pet cats wear collars. Billie randomly samples 100 pet cats to see whether they are wearing collars. Can Billie use the normal approximation?

2 Suppose that 5 percent of the time, a GPS map contains an error. You want to randomly sample 50 GPS maps to see how many contain errors. Can you use the normal approximation?

3 Suppose that 10 percent of L.E.D. TVs cost over $500.00. TV Land has hundreds of TVs. Bob wants to randomly sample n TVs and count the number of TVs that cost over $500.00. How many TVs does he need to sample in order to use the normal approximation? (Use the conditions to solve for n.)

4 Carey wants to sample baseball players in her league to see how many of them wear Acme cleats. It is known that about 50 percent of players wear Acmes. How many players does Carey need to sample to use the normal approximation?

5 Rafael looks for typos on random pages of a book. He samples 150 pages of a book. The probability of an error on a page is 0.01. Is there enough data to do a normal approximation?

Forming Z Using the Mean and Standard Deviation of the Binomial

In this section, you see how to form the Z statistic using parts of the binomial distribution as discussed earlier. When you use the normal distribution, you find $Z = \frac{X - \mu}{\sigma}$ (see Chapter 9). When you approximate the binomial distribution with the normal, you use the mean and standard deviation of the binomial for μ and σ. The mean is np, and the standard deviation is $\sqrt{np(1-p)}$, where n is the sample size and p is the probability of yes (success). So, the Z value looks like this: $Z = \frac{X - np}{\sqrt{np(1-p)}}$.

This formula may look difficult, but it only contains n, p, and the value of X that you need!

TIP

EXAMPLE

Q. Suppose that 70 percent of pet cats wear collars. Billie randomly samples 100 pet cats to see whether they are wearing collars. She finds that 65 are wearing collars. What does the Z statistic for the normal distribution look like?

A. First check the conditions: np = 100(0.70) = 70 and 100(1 − 0.70) = 30 are both at least 10. Therefore,

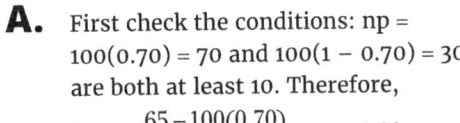

$$Z = \frac{65 - 100(0.70)}{\sqrt{100(0.70)\,(1 - 0.70)}} = -1.09.$$

 6   Suppose that 10 percent of L.E.D. TVs cost over $500.00. TV Land has hundreds of TVs. Bob wants to randomly sample 100 TVs. You want the chance that at most 15 of them are over $500.00. What is your Z value?

 7   Suppose that 80 percent of all U.S. stamps used are "Forever Stamps," which means the postage doesn't change from what was paid for the stamp. Andre randomly samples 100 stamps. He wants to know the chance that at least 85 of them are Forever Stamps. What does his Z statistic look like?

8. Maryanna cuts hair. She knows that 70 percent of her clients want a haircut when they come into her salon. She randomly samples 200 clients. She wants to know the chance that at least 150 of her clients want a haircut. What is the Z value?

9. Carol knows that 25 percent of her tomatoes can be sold. She samples 40 tomatoes at random. What's the Z value if she wants to know the chance that fewer than ten can be sold?

10. Rafael looks for errors in books and samples 1,000 pages at random to review. The chance that a page has an error is 1.0 percent.

   a. Can he use the normal approximation?

   b. Suppose that he finds 20 errors. What is the Z value?

Approximating to Solve Large-Scale Probability Problems

Once you have checked your conditions and found Z, you will then find the probability for Z by looking up the Z-value on the Z-table (Table A-2 in the appendix), finding the probability being asked for, and then answering the question. Remember that the Z-table shows you the probability of being less than or equal to the value you look up. (Flip to Chapter 9 for more on using the Z-table.)

EXAMPLE

Q. Suppose that 70 percent of pet cats wear collars. Billie samples 100 pet cats to see whether they are wearing collars. What's the chance that at least 65 cats are wearing collars?

A. The conditions check out; you have
$$Z = \frac{65 - 100(0.70)}{\sqrt{100(0.70)(1-0.70)}} = -1.09.$$
You want $P(X \geq 65) = P(Z \geq -1.09) = 1 - 0.1379 = 0.8621.$

11 Suppose that 10 percent of L.E.D. TVs cost over $500.00. TV Land has hundreds of TVs. Bob wants to randomly sample 100 TVs. You want the chance that at most 15 of them are over $500.00. Find this probability.

12 Suppose that 80 percent of all U.S. stamps used are "Forever Stamps," which means the postage doesn't change from what was paid for the stamp. Andre randomly samples 100 stamps. He wants to know the chance that at least 85 of them are Forever Stamps. Find this probability.

13. Maryanna cuts hair. She knows that 70 percent of her clients want a haircut when they come in. She wants to know the chance that at least 75 of her clients want a haircut out of a random sample of 100. Find this probability.

14. A bookstore knows that 10 percent of all books contain an error in the first chapter. You randomly sample 100 books. What's the chance that more than 11 contain an error in the first chapter?

15. Rafael looks for errors in books and samples 1,000 pages at random to review. The chance that a page has an error is 1.0 percent. What's the chance he finds more than 20 errors?

Solutions to Problems in Approximating a Binomial with a Normal Distribution

(1) $p = 0.70$ and $n = 100$; therefore, $np = 100(0.70) = 70$ and $n(1 - p) = 100(1 - 0.30) = 30$ are both at least 10, so yes, Billie can use the normal approximation.

(2) $p = 0.05$ and $n = 50$; therefore, $np = 50(0.05) = 2.5$ is not at least 10, so no, you can't use the normal approximation.

(3) You have n and $p = 0.10$. You need to meet two conditions: (1) np needs to be at least 10, and (2) n(1-p) needs to be at least 10. First, np needs to be at least 10, so n(0.10) is at least 10, and that means n is at least $10/0.10 = 100$ for the first condition. For the second condition, $n(1 - p)$ needs to be at least 10, or $n(1 - 0.10) = n(0.90)$ to be at least 10. This means n must be at least $10/.90 = 11.11$, which means you need at least 12 for the second condition. Both conditions must be met, so n needs to be at least 100.

(4) To determine how many players Carey needs to sample to use the normal approximation, $np = n(0.50)$ needs to be at least 10, so n has to be at least 20. Same for $1 - p = 0.50$. So n needs to be at least 20.

(5) You have $n = 150$ and $p = 0.01$. You know $np = 150(0.01) = 1.50$, which does not meet the first condition. Therefore, you cannot use the normal approximation. Note that p is so small that the data would be very skewed, with not many yeses and a whole lot of nos. It would not be balanced like a normal distribution.

(6) $p = 0.10$ and $n = 100$, so the conditions are met, and $Z = \dfrac{X - np}{\sqrt{np(1-p)}} = \dfrac{15 - 100(0.10)}{\sqrt{100(0.10)(1-0.10)}} = 1.67$.

(7) $n = 100$, $p = 0.80$, and $1 - p = 0.20$, so the conditions are met, and $Z = \dfrac{X - np}{\sqrt{np(1-p)}} = \dfrac{85 - 100(0.80)}{\sqrt{100(0.80)(1-0.80)}} = 1.25$.

(8) The Z value is $Z = \dfrac{150 - 200(0.70)}{\sqrt{200(0.70)(0.30)}} = 1.54$.

(9) $p = 0.25$ and $n = 40$. The Z value is found by taking $P\left(Z < \dfrac{10 - 40(0.25)}{\sqrt{40(0.25)(0.75)}} \right) = P(Z < 0)$. So, the value of Z is zero.

(10) Rafael looks for errors in books and samples 1,000 pages at random to review. The chance that a book has an error is 1.0 percent.

    a. To determine whether Rafael can use the normal approximation with this information, you first check $np = 1,000(0.01) = 10$, which meets the conditions. Then you check $n(1 - p) = 1,000(0.99) = 990$, which more than meets the conditions. So the answer is yes, Rafael can use the normal approximation.

    b. Rafael finds $X = 20$ errors. His Z value is $Z = \dfrac{20 - np}{\sqrt{np(1-p)}} = \dfrac{20 - 1000(0.10)}{\sqrt{1.000(0.01)(0.99)}} = \dfrac{10}{3.146} = 3.18$.

(11) The probability is $P\left(Z \le \dfrac{15-100(0.10)}{\sqrt{100(0.10)(0.90)}}\right) = P(Z \le 1.67) = 0.9525$.

(12) The probability is $P\left(Z \ge \dfrac{X-np}{\sqrt{np(1-p)}}\right) = P\left(Z \ge \dfrac{85-100(0.80)}{\sqrt{100(0.80)(1-0.80)}}\right) = P(Z \ge 1.25) =$

$1-.8944 = 0.1056$.

(13) The probability is $P\left(Z \ge \dfrac{75-100(0.70)}{\sqrt{100(0.70)(1-0.70)}}\right) = P\left(Z \ge \dfrac{5}{4.58}\right) = P(Z \ge 1.09) = 1-.8621 = 0.1379$.

(14) $n = 100$ and $p = 0.10$. The probability is $P(X > 11) = P\left(Z > \dfrac{11-100(0.10)}{\sqrt{100(0.10)(1-0.10)}}\right) = P(Z > 0.33) =$

$1-0.6293 = 0.3707$.

(15) $n = 1{,}000$ and $p = 0.01$. The chance Rafael finds more than 20 errors is

$P(X > 20) = P\left(Z > \dfrac{20-np}{\sqrt{np(1-p)}}\right) = P\left(Z > \dfrac{20-1000(0.01)}{\sqrt{1{,}000(0.01)\,(0.99)}}\right) =$

$P\left(Z > \dfrac{10}{3.146}\right) = P(Z > 3.18) = 1-0.9993 = 0.0007$.

Chapter **11**

Sampling Distributions and the Central Limit Theorem

A *sampling distribution* is a special distribution made from sample statistics. The sample statistics can be the sample means (averages) or the sample proportions (proportion in the sample with a certain characteristic), for example. They are important because if you are trying to estimate or test a hypothesis about a population parameter (such as the population mean or population proportion), you need to know what the distribution is for all possible results you could get, what the mean is, what the standard deviation is, and how to use it to find probabilities about your own sample mean or proportion.

For example, the sampling distribution of the sample mean is the set of all possible sample means taken from all possible samples of size n from your population. The sampling distribution of the sample proportion is the set of all possible sample proportions taken from all possible samples of size n from your population, where the proportion is the number of individuals in the sample with your characteristic divided by the sample size.

In this chapter, you find out what a sampling distribution does and how it is used to find probabilities for sample statistics. This is important for later chapters involving hypothesis testing and confidence intervals.

Surveying a Sampling Distribution

The sampling distribution of the sample mean is the collection of all possible sample means from all possible samples of size n (fixed) taken from the population. The sample mean in the most general sense is a random variable that changes value from sample to sample. It is written as $\bar{X}$. It has a distribution, a mean, a standard deviation (which is called the *standard error*), and it can often be transformed to the standard normal (Z) distribution to find probabilities for the sample mean. When you take your own sample mean, you can use this information to find the probability that your sample mean is greater than, less than, or between certain values.

For example, suppose that you want to study the average gas price in the United States. To do so, you take a random sample of gas prices, say from 100 gas stations, and find the sample mean of those 100 prices. But to know more about what the average gas price is for all U.S. gas stations, you need to know more about how the sample mean behaves.

EXAMPLE

Q. Tell whether the following statement is true or false: "The sampling distribution of the sample mean is the group of all statistics of any type and size that come from the population."

A. False. The sampling distribution of the sample mean is the group of all sample *means* (only) that come from all samples of size n (*fixed*) from the population.

---

1 Tell whether the following statement is true or false: "The sampling distribution of $\bar{X}$ has its own mean and standard deviation (called the standard error)."

2 Tell whether the following statement is true or false: "$\bar{X}$ has its own distribution of values and can be transformed to the Z distribution in many cases to find probabilities."

③ Tell whether the following statement is true or false: "$\bar{X}$ is a random variable."

Examining the Mean and the Standard Error of $\bar{X}$

The mean of the sampling distribution of $\bar{X}$ has the same value as the mean of the population from which the samples were drawn. The mean of $\bar{X}$ is $\mu_{\bar{x}}$, and the mean of the original population is μ_X. And they have the same value. The mean of all the sample means from the population is the same as the population's mean.

The standard deviation of $\bar{X}$ is called the *standard error of* $\bar{X}$. You get the standard error by dividing the population standard deviation by the square root of n (because you are averaging, n goes in the denominator). The standard error of $\bar{X}$ is written as $\sigma_{\bar{x}}$ and is calculated by $\frac{\sigma_X}{\sqrt{n}}$. Notice that as the variability in the population, σ_X, increases, the variability in the sample means increases (as the population gets more varied, so will the means drawn from that population). And as the sample size n increases, the variability in the sample means goes down (more data, more precision).

EXAMPLE

Q. Tell whether the following statement is true or false: "As the sample size increases, the mean of $\bar{X}$ increases."

A. False. The mean of $\bar{X}$ is the same as the mean of the population of X, no matter what.

4 Tell whether the following statement is true or false: "The mean of $\bar{X}$ is equal to the value μ_X."

5 Tell whether the following statement is true or false: "The standard error of $\bar{X}$ has the same value as the standard deviation of the population."

6 Tell whether the following statement is true or false: "The standard error of $\bar{X}$ increases as n increases."

7 Fill in the blank with the best answer: The standard error of $\bar{X}$ _____ as the standard deviation of the population increases.

Exploring the Shape of $\bar{X}$ and the Central Limit Theorem

The shape of the sampling distribution of the sample mean $\bar{X}$ is normal if the original population is normal. In other words, the general shape of the population of sample means stays the same as the general shape of the original population if the population is normal. It has the bell shape and the same mean, μ_X, but the standard error formula, $\sigma_X/\sqrt{n}$, is smaller because you are averaging the values from the samples.

However, if the original population is not normal, something perhaps surprising happens — the sampling distribution of the sample mean $\bar{X}$ is approximately normal if the size of your sample (n) is large enough. Typically, 30 will do, although the important part of this story is that the larger the sample you take, the more the sampling distribution of $\bar{X}$ looks like a normal distribution. Again, it has the same mean, μ_X, and it has the same standard error formula, $\sigma_X/\sqrt{n}$, since you are averaging the values from the samples. This result is called the *Central Limit Theorem.*

EXAMPLE

Q. Tell whether the following statement is true or false: "The Central Limit Theorem only applies when the original population is not normal."

A. True. The Central Limit Theorem is not needed if the original population is normal.

---

8 The Central Limit Theorem is primarily focused on which of the following aspects of the sampling distribution of $\bar{X}$?

   a. The center (the mean)

   b. The variability (standard error)

   c. The shape

What tends to be true about the sampling distribution of $\bar{X}$ if X is not normal? Choose the right answer.

a. The shape is exactly normal if n > 30.

b. The shape is approximately normal if n > 30 and gets more normal as n increases.

c. The shape is approximately normal for any n and gets more normal as n increases.

d. The shape stays the same as the shape of the original population X.

 Tell whether the following statement is true or false: "The mean and standard error increase toward the mean and standard deviation of the original population X as n increases."

11 Tell whether the following statement is true or false: "If X is normal, you need n > 30 to find probabilities for $\bar{X}$."

12　Tell whether the following statement is true or false: "n > 30 is a magic number. If n is 30, the Central Limit Theorem would never work."

Finding Probabilities for the Sample Mean

Now that you know the major characteristics of the sampling distribution of the sample mean $\bar{X}$, you are able to find probabilities for it. If X has a normal distribution, you know that $\bar{X}$ has a normal distribution, which means in order to find probabilities for $\bar{X}$, you transform $\bar{X}$ to a Z-value using the formula, $Z = \dfrac{\bar{X} - \mu_{\bar{X}}}{\sigma_{\bar{X}}} = \dfrac{\bar{X} - \mu_X}{\sigma_X / \sqrt{n}}$.

You also know by the Central Limit Theorem that if X does not have a normal distribution but n is large enough, typically greater than 30, you can use the same formula to transform $\bar{X}$ to a Z-value, $Z = \dfrac{\bar{X} - \mu_{\bar{X}}}{\sigma_{\bar{X}}} = \dfrac{\bar{X} - \mu_X}{\sigma_X / \sqrt{n}}$, and find probabilities for it.

In both cases, you use the Z-transformation to find probabilities for sample means and see how typical or atypical they might be under the circumstances.

For example, suppose that test scores have a normal distribution with a mean of 80 and a standard deviation of 5. What's the chance that ten randomly chosen test scores are higher than 85? To determine this, you first check to see if you have a normal distribution, which you do — so now you don't need n to be 30; $\bar{X}$ is normal for any sample size. Then you transform to Z using the formula, $Z = \dfrac{\bar{X} - \mu_{\bar{X}}}{\sigma_{\bar{X}}} = \dfrac{\bar{X} - \mu_X}{\sigma_X / \sqrt{n}} = \dfrac{85 - 80}{5 / \sqrt{10}} = 3.16$. From here, you want $P(X > 85) = P(Z > 3.16) = 1 - 0.9992 = 0.0008$ according to the Z-table (Table A-2 in the appendix). This means it's pretty atypical (unusual) to have ten exam scores average higher than 85 when the population average is 80.

EXAMPLE

Q. Answer yes or no to the following question: Suppose that your test scores have a skewed (non-normal) distribution. Would you still be able to find $P(\bar{X} > 80)$ using the same Z-transformation from the previous example?

A. No. The solution would require the Central Limit Theorem and n > 30.

13 Suppose that men's heights have a normal distribution with a mean of 72 inches and a standard deviation of 4 inches. You take a random sample of 36 men.

   a. What is the chance that men's average height is under 70 inches?

   b. Did you need the Central Limit Theorem to solve the previous problem using the Z-transformation?

   c. Did n have to be at least 30 in the previous problem in order to solve it using the Z-transformation?

14 Suppose that the amount of time it takes employees to finish a specific task has a skewed-right distribution with a mean of 15 minutes and a standard deviation of 2 minutes.

   a. What is the chance that 36 randomly chosen employees take more than 17 minutes to do the task?

   b. Can you take samples of any size to do the Z-transformation to find probabilities for the average time to finish the task? Explain your answer.

   c. Did you need the Central Limit Theorem to do the problem in part a? Why or why not?

   d. How would your answer to part a change if the time to finish the tasks had a normal distribution? Explain.

Suppose that you own a Mexican food truck. The time it takes you to fill an order has a normal distribution with a mean of 5 minutes and a standard deviation of 1 minute.

a. With a random sample of 16 orders, what is the chance that it takes you less than 5½ minutes to fill an order?

b. Did n need to be > 30 here to use Z?

c. Did you need to use the Central Limit Theorem to use the Z-transformation?

 Exam scores have an unknown distribution with a mean of 85 and a standard deviation of 5.

a. What is the chance that 49 randomly chosen exams have a mean that is lower than 83?

b. Did you need n to be > 30 here in order to use Z?

c. Did you need to use the Central Limit Theorem to use the Z-transformation?

Solutions to Problems in Sampling Distributions and the Central Limit Theorem

(1) True. The sampling distribution of $\bar{X}$ has its own mean and standard deviation (called the standard error) because it is a random variable.

(2) True. $\bar{X}$ has its own distribution of values and can be transformed to the Z distribution in many cases to find probabilities. If X doesn't have a normal distribution, but the sample size is large enough, the Z-transformation can be used. If X has a normal distribution, the Z-transformation can always be used.

(3) True. $\bar{X}$ is a random variable. It takes on random values each time you select a different sample from the population.

(4) True. The mean of $\bar{X}$ is μ_X. It's notation is $\mu_{\bar{x}}$ and its value is μ_X.

(5) False. The standard error of $\bar{X}$ is the standard deviation of the population divided by the square root of n.

(6) False. The standard error of $\bar{X}$ does not increase as n increases; it decreases.

(7) The standard error of $\bar{X}$ INCREASES as the standard deviation of the population increases.

(8) c., the shape.

(9) b., the shape is approximately normal for n > 30 and gets more normal as n increases.

(10) False. The mean and standard error do not increase toward the mean and standard deviation of the original population X as n increases. The mean of the sampling distribution stays the same as the mean of X, and the standard error is the population standard deviation divided by the square root of n.

(11) False. If X is normal, you do not need n > 30 to find probabilities for $\bar{X}$.

(12) False. The value of n > 30 is not a magic number; it's just a threshold, or holding place. If n is 30, the Central Limit Theorem could still work; it just works better and better as n increases. It depends on how far the original distribution is away from the normal distribution.

(13) a. You want $P(\bar{X} < 70)$ $Z = P\left(Z < \dfrac{70-72}{4/\sqrt{36}}\right) = P(Z < -3) = 0.0013$.

   b. No. You do not need the Central Limit Theorem to do the previous problem using the Z-transformation; X already had a normal distribution.

   c. No. You do not need n to be at least 30 in the previous problem in order to do it using the Z-transformation. X already had a normal distribution.

(14) a. You want $P(\bar{X} > 17) = P\left(Z < \dfrac{17-15}{4/\sqrt{36}}\right) = P(Z > -3) = 0.0013.$

b. No. You cannot take samples of any size to do the Z-transformation to find probabilities for the average time to finish the task because X doesn't have a normal distribution. You need n > 30.

c. Yes. You did need the Central Limit Theorem to do the problem in part a because X did not have a normal distribution. It had a skewed distribution.

d. It's an approximation, but the answer would not change. The calculations are the same; you just have to check the condition that n > 30 in the original problem, but not if X had a normal distribution.

(15) a. You want $P(\bar{X} < 5.5) = P\left(Z < \dfrac{5.5-5}{1/\sqrt{16}}\right) = P(Z < 2) = 0.9772.$

b. No. n didn't have to be > 30 in this situation to use the Z-transformation. X already had a normal distribution.

c. No. You did not need to use the Central Limit Theorem to use the Z-transformation.

(16) a. You want $P(\bar{X} < 83) = P\left(Z < \dfrac{83-85}{5/\sqrt{49}}\right) = P(Z < -2.8) = 0.0226.$

b. Yes. You need n to be > 30 to use the Z-transformation. The distribution of X was unknown.

c. Yes. You need the Central Limit Theorem because you didn't know the distribution of X. (In other words, you couldn't say that it was normal.)

Chapter **12**

Probability's Role in Confidence Intervals and Hypothesis Tests

I n this chapter, you explore how probability plays a role in two major types of *inferences* in statistics (where you move from sample results to generalizations about the population): confidence intervals and hypothesis tests.

A confidence interval is used when you have a population parameter (a value, usually unknown, that summarizes the population, such as the population mean or population proportion of yeses) and you want to estimate it. For example, what is the average price of gas in the United States?

A hypothesis test is a test about a population parameter value, such as the population mean or proportion of yeses. For example, is the mean gas price in the United States equal to a certain value, or is it higher than that? Probability is involved in both situations; this chapter shows you how.

Reviewing Confidence Intervals and Probability

Probability is present in nearly every topic in statistics, but none more so than the two types of inference: confidence intervals and hypothesis tests. In this section, you see where probabilities come into play regarding confidence intervals.

REMEMBER

A *confidence interval* is a range of likely values for a population parameter, such as the population mean. In the case of the population mean, the interval is the sample mean plus or minus a margin of error, which is made up of a certain number of standard errors — the number depends on how confident you want to be with your interval. If you want to be 95 percent confident, you add and subtract about two standard errors on each side of the sample mean; if you want to be 99 percent confident, you add and subtract about three standard errors. For the full details on confidence intervals, see *Statistics For Dummies* (also written by yours truly and published by Wiley).

The exact number of standard errors to add and subtract is actually a Z value from the Z table (Table A-2) in the appendix, which depends on how confident you want to be. A 95 percent confidence level means Z is about 2 or, more specifically, 1.96. So the formula for a 95 percent confidence interval for the population mean is $\bar{X} \pm 1.96 \frac{\sigma}{\sqrt{n}}$.

For example, say you want to estimate the mean price of gas in the United States. You sample 100 random gas stations and you find that the mean gas price is $3.20. Suppose that the population standard deviation is 20 cents ($0.20). Your 95 percent confidence interval for the mean price of a gallon of gas in the United States is $3.20 \pm 1.96 \frac{0.20}{\sqrt{100}} = 3.20 \pm 0.04 = (\$3.16, \$3.24)$. You say you are 95 percent confident that the mean gas price for the United States is in this range.

The 0.04 in the previous example represents the margin of error; it's the total amount that is added to and subtracted from the mean in a confidence interval. It's made up of the Z value (in the previous example, the Z value was 1.96) and the standard error, which is made up of the standard deviation divided by the square root of n. As n increases, the margin of error decreases — more good data leads to more precision. As the standard deviation, σ, increases, the margin of error increases — more variability in the population leads to more variability in your sample mean.

And finally, if you increase the confidence level from 95 percent to 99 percent — say you'll be adding and subtracting more standard errors (from about 2 to about 3) — your confidence interval widens as the margin of error increases.

Where does the probability come in? It's related to the confidence level. If you have a 95 percent confidence level in your confidence interval, that doesn't mean there is a 95 percent probability that the real average gas price in the United States is in that range — once you collect the data, the value is in your confidence interval or it isn't. That means if you repeat the whole process over and over again, many, many (technically infinite) times, collecting data and making a confidence interval each time, 95 percent of the intervals should contain the real population mean (and 5 percent shouldn't).

The thing is, you never know what the real population mean is, so you don't know which ones are right or wrong. Plus, you only make one interval in real life, so it's just a confidence level that you use when describing it, not an actual probability. You'd say, "We are 95 percent confident that the population mean is in this range."

If you increase the confidence level from 95 percent to 99 percent, the percentage of intervals containing the true population parameter also increases, as the wider the interval is, the easier it is to capture the population parameter inside of it. For the previous price of gas example, the original 95 percent confidence interval was $3.20 plus or minus $0.04 for the average price of a gallon of gas. If you want to be 99 percent confident, the Z value would become about 3 (actually 2.57), and the confidence interval would be wider: $3.20 \pm 2.57 \dfrac{0.20}{\sqrt{100}} = 3.20 \pm 0.5 = (\$3.15, \$3.25)$.

That means with an infinite number of repetitions of this process of collecting new data and finding new confidence intervals, you would find that 99 percent of them would contain the actual population mean gas price in the United States. You just hope yours is one of them!

EXAMPLE

Q. Tell whether the following statement is true or false: "You interpret a confidence interval by stating the probability that the population parameter is in the interval."

A. False. The confidence level represents the percentage of times that infinitely repeated confidence intervals contain the true population parameter.

---

1 Tell whether the following statement is true or false: "You know whether the population parameter is in your confidence interval after you collect your data."

2 Tell whether the following statement is true or false: "95 percent confidence means there is a 95 percent probability that your parameter is in your confidence interval."

3 Tell whether the following statement is true or false: "If you repeat the process of data collection and make 100 confidence intervals, exactly 95 of them will contain the true value of the parameter."

4 Fill in the blank with the best answer: If you decrease the confidence level, the confidence interval gets _____. This also decreases the confidence you have in your results.

5 Fill in the blank with the best answer: If you decrease the sample size, the confidence interval gets _____.

6 Fill in the blank with the best answer: If the sample standard deviation were to decrease, the confidence interval gets

_____.

7 Tell whether the following statement is true or false: "A confidence interval is a range of likely values for your sample statistic, like the sample mean."

 8 Tell whether the following statement is true or false: "You never know if your confidence interval contains the true population parameter. It either does or it doesn't."

Exploring Hypothesis Testing and Probability

A *hypothesis test* is a statistical technique in which you test a hypothesized value of the population parameter and use your data to determine whether the data causes you to reject the hypothesized value or fail to reject it. Your data is your evidence. Probability comes into play when you weigh the evidence to see how far away the sample is from the hypothesis about the population. That number is called a p-value. (For full details about hypothesis tests and p-values, see *Statistics For Dummies*.)

For example, say you want to test whether the average fill amount in yogurt cups is really at the stated amount of 8 ounces; you believe it's less than that. You take a random sample of 36 filled cups and weigh each of their fill amounts. You find that the average fill amount is 7.8 ounces. The population standard deviation is 1 ounce. What do you decide?

Your null and alternative hypotheses are $H_o : \mu = 8$ versus $H_a : \mu < 8$. Your test statistic is $Z = \dfrac{\bar{X} - \mu_o}{\sigma / \sqrt{n}} = \dfrac{7.8 - 8}{1 / \sqrt{36}} = -1.2$. This value on the Z table is not far from zero, which means 7.8 isn't that far below 8, but how much below is it? That's where the probability comes in.

REMEMBER
To measure the difference between the data you got and the null hypothesis you have, you use the p-value. A *p-value* is a probability that measures how much evidence you have against the null hypothesis. If the p-value is large, you can't reject the null hypothesis because the results you got were the results you were supposed to get, according to H_o.

If the p-value is small enough, you can reject the null hypothesis because the results you got were results you only have a small chance of getting according to H_o. The farther the Z value is out on the tail of the Z distribution (in this case, the left tail because it is negative), the smaller the p-value gets, and the farther you get from H_o. Looking up –1.2, our test statistic, in the Z table to measure this left-tail probability, you get 0.1151. This is the chance of being –1.2 standard deviations below H_o or more.

Because 0.1151 is a fairly large number in terms of probability, you fail to reject H_o. You set up your cutoff points ahead of time, but even a large cutoff point of 0.100 is not met here. So you fail to reject H_o and conclude that the yogurt cups are not being underfilled. In other words, you can't claim the alternative hypothesis, H_a, is true.

EXAMPLE

Q. Tell whether the following statement is true or false: "The p-value is the probability that H_o is true."

A. False. The p-value is the chance of getting the test statistic you got, or results that are further from H_o, given H_o is true.

---

9 Tell whether the following statement is true or false: "If the p-value is small enough, we reject the H_a and accept the H_o."

10 Tell whether the following statement is true or false: "If the p-value is small enough, we have strong evidence against H_o."

11. Tell whether the following statement is true or false: "If the p-value is smaller than our cutoff point (our significance level), we reject H_o."

12. Tell whether the following statement is true or false: "The p-value is the probability that H_a is true."

13. Tell whether the following statement is true or false: "If you were to collect a new data set and find a new test statistic, your p-value would also change."

14. Suppose that $H_o : \mu = 8$ versus $H_a : \mu > 8$ and your test statistic is 1.2. What is your p-value?

15 Suppose that $H_o : \mu = 8$ versus $H_a : \mu \neq 8$ and your test statistic is 1.2. What is your p-value?

Solutions to Problems in Probability's Role in Confidence Intervals and Hypothesis Tests

1. False. You never know whether the population parameter is in your confidence interval after you collect your data. It is either in the interval, or it isn't. You can only say you are 95 percent confident if it is a 95 percent confidence interval, for example.

2. False. Ninety-five percent confidence does not mean there is a 95 percent probability that your parameter is in your confidence interval. The 95 percent is for the long-term probability after (infinitely) many intervals have been performed (which is in theory only). It doesn't apply to a single interval.

3. False. If you repeat the process of data collection and make 100 confidence intervals, exactly 95 of them might contain the population parameter, but other values could contain the parameter too. One hundred is not enough to say that exactly 95 will be in there.

4. If you decrease the confidence level, the confidence interval gets NARROWER. This also decreases the confidence you have in your results.

5. If you decrease the sample size, the confidence interval gets WIDER.

6. If the sample standard deviation were to decrease, the confidence interval gets NARROWER. This is because the variability in the population goes down, so the variability in the means goes down.

7. False. A confidence interval is a range of likely values for your *population parameter*. It is based on the sample mean.

8. True. You never know whether the parameter is in the interval you made; that's why you use the confidence level.

9. False. If the p-value is small enough, you reject the H_o. You conclude that H_a is true.

10. True. If the p-value is small enough, you have strong evidence against H_o. The test statistic is far out into the tail, and the sample mean is far from the hypothesized value.

11. True. The significance level cuts off where you reject and where you fail to reject.

12. False. The p-value is not the probability that H_a is true. H_a is either true or it isn't. The probability is about your test statistic and how likely it is, given that H_o is true.

13. True. P-values change every time the test statistic changes.

14. The p-value is the same as in the example, 0.1151. It's just in the right tail instead of the left tail.

15. In this case, the p-value doubles to account for the two tails you could have been in due to the not-equal-to sign. So it is $0.115 \times 2 = 0.2302$.

4

Taking It up a Notch: Advanced Probability Models

IN THIS PART . . .

Identify the characteristics of the Poisson distribution and how it differs from a binomial distribution.

Find probabilities for the geometric distribution and calculate its mean, variance, and standard deviation.

Explore the negative binomial distribution and find probabilities for it.

Measure the hypergeometric distribution's mean, variance, and standard deviation.

Chapter **13**

Working with the Poisson (a Nonpoisonous) Distribution

The *Poisson distribution*, pronounced "pwa-san" is an interesting distribution with lots of uses. In this chapter, you explore the characteristics of the Poisson distribution, and the mean, standard deviation, variance, and probabilities for it. You see what happens when you change the units of a Poisson distribution, and how the normal distribution can approximate the Poisson distribution in large cases.

Identifying a Poisson Distribution from a Binomial Distribution

The binomial distribution (discussed in Chapter 8) has a fixed number of n trials, and for each trial, it has two possible outcomes: yes and no (or success and failure). The binomial counts the number of yeses among those fixed trials, X, and the values of X go from $0, 1, 2, \ldots, n$. A Poisson distribution involves a fixed period of time or space, and you are counting the number of occurrences within that period of time or space, X. The values of X go from $0, 1, 2, \ldots$. There is no end to the possible whole numbers that X can take on.

The Poisson distribution has two other important characteristics:

>> The events occur independently of each other.

>> No two events occur at the same time.

For example, while the binomial distribution may count whether a car rolls through the stop sign at an intersection and totals up the number of cars out of 100 that do roll through the stop sign, the Poisson distribution counts the number of cars that roll through the stop sign in an hour, a day, a week, or however long.

A Poisson distribution also involves a fixed amount of space. For example, suppose that X is the number of chocolate chips in a 2-inch diameter chocolate chip cookie. You may want to know the probabilities that X is small, indicating the cookie doesn't have enough chocolate chips, or that X is large, indicating that the cookie has too many chocolate chips.

REMEMBER

The binomial is yes or no, trial by trial, for n fixed trials. The Poisson distribution counts the number of occurrences in a fixed period of time or space. It's not a yes or no issue.

EXAMPLE

Q. Binomial or Poisson? You are counting the number of accidents that occur at a certain intersection in one month.

A. Poisson. You are fixing time, not the number of trials, and you are counting the number of occurrences that happen during that fixed period of time.

1. Bernardo counts the number of phone calls that go to voicemail during his 1-hour lunch break. Explain why this is a Poisson distribution.

2. What numerical values can X take on with a Poisson distribution?

3. Name two differences between the binomial distribution and the Poisson distribution.

Obtaining Probabilities for a Poisson Distribution

The Poisson probability mass function (pmf) is the function that gives us the probabilities for a Poisson distribution. The pmf for the Poisson distribution is $\dfrac{e^{-\lambda}\lambda^x}{x!}$ where $x = 0, 1, 2, 3,$ and so on.

As x gets larger and larger, the probabilities get smaller and smaller for all values of λ. The value of e is on your calculator and is an irrational number (about 2.72). x! is $x(x-1)\ldots(3)(2)(1)$. Lamda (λ) is the parameter for the Poisson distribution; it can change with each problem.

You can find single probabilities using the pmf formula. For example, suppose that an ice cream shop owner wants to count the number of ice cream cones sold per hour (X). X is a Poisson distribution with $\lambda = 10$. The owner wants to know the probability of selling 12 cones within 1 hour, so you find $\dfrac{e^{-10}10^{12}}{12!} = \dfrac{e^{-10}10^{12}}{12!} = 0.095$.

You can also use the pmf formula to find the number of blemishes in 1 square yard of carpet. Suppose that $\lambda = 0.2$ and you want to know P(X = 1). You would find $\dfrac{e^{-0.2}0.2^1}{1!} = 0.164$.

Table A-3 in the appendix includes a Poisson table that finds probabilities for $X \le x$ for various values of λ. For example, if you want $P(X \le 12)$ where X is a Poisson distribution with $\lambda = 10$, you would go to Table A-3 and intersect $\lambda = 10$ (in the column) with 12 (in the row), and you get 0.792. This is a large probability because it sums all the probabilities from P(X = 0) to P(X = 12).

If you want P(X < x), then you subtract one value from x and find $P(X \le x - 1)$. For example, if you want P(X < 12) you rewrite it as $P(X \le 11)$ and then look up $\lambda = 10$ and X = 11 in Table A-3 to get 0.697.

If you want $P(X \ge 12)$, rewrite it as $1 - P(X \le 11)$ as the complement, and use the table. You know from the previous example that $P(X \le 11) = 0.697$, so the answer is $1 - 0.697 = 0.303$.

If you want P(X > 12), rewrite it as $1 - P(X \le 12)$. You know $P(X \le 12) = 0.792$, so $P(X > 12) = 1 - 0.792 = 0.208$.

TIP

To find probabilities between two values, rewrite them if needed so they are in terms of $\le$ and subtract. For example, $P(9 \le X < 12)$ is rewritten as $P(X \le 11) - P(X \le 8)$. For these problems, a number line really helps. See Figure 13-1.

FIGURE 13-1:
$P(9 \le X < 12) =$
$P(X \le 11) -$
$P(X \le 8)$
on a number line.

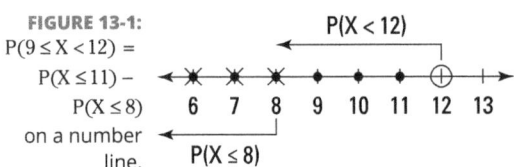

EXAMPLE

Q. The owner of a car wash wants to know how many people go through the car wash in 1 hour. X is the number of customers, and is a Poisson distribution with $\lambda = 5$. What's the chance that X = 7?

A. $\frac{e^{-5}5^7}{7!} = 0.105$. There is a 0.105 percent chance that X = 7.

---

4 Continuing with the car wash example, what's the chance that more than seven customers come through the car wash in 1 hour?

5 What's the chance that fewer than six customers come through the car wash in 1 hour?

---

6 What's the chance that X is between 5 and 8, inclusive?

Finding the Mean, Variance, and Standard Deviation of the Poisson Distribution

The Poisson distribution is interesting in that the mean is λ, the variance is also λ, and the standard deviation is the square root of λ.

For example, if X is the number of planes that land at a city airport in 1 hour and is a Poisson distribution with $\lambda = 5$ arrivals per hour, then the mean is 5, the variance is 5, and the standard deviation is $\sqrt{5}$.

If you have the following pmf for another Poisson distribution, $\dfrac{e^{-5}5^x}{x!}$, you know λ, so the mean and variance are 5, and the standard deviation is $\sqrt{5}$.

TIP

If you are doing a quality control experiment and you notice that the mean of the Poisson distribution is too large, say you have on average too many chocolate chips in your cookies (is there such a thing?), you can make adjustments to lower the value of the average, which is λ. By doing so, you get the added bonus of lowering the variance as well. So in this example, there are fewer chocolate chips on average, but also the number of chips in each cookie doesn't vary as much.

EXAMPLE

Q. Write down the pmf of the Poisson distribution whose standard deviation is 3.

A. The standard deviation is $\sqrt{\lambda} = 3$, so $\lambda = 9$, and the pmf is $\dfrac{e^{-9}9^x}{x!}$.

 7 Suppose that the mean of a Poisson distribution is 4. What is the variance and standard deviation?

 8 What is the variance and standard deviation of the Poisson distribution with a pmf of $\dfrac{e^{-16}16^x}{x!}$?

9 Tell whether the following statement is true or false: "The mean and standard deviation of the Poisson distribution are the same."

Processing the Poisson Process: The Business of Changing Units

Another way in which the Poisson distribution is interesting is that when you change the units of time and space by multiplying by t, the number of occurrences stays a Poisson distribution: the mean is λt, the variance is also λt, and the standard deviation is the square root of λt.

For example, if X is the number of planes that land at a city airport in 1 hour and is a Poisson distribution with $\lambda = 5$ arrivals per hour, then the mean is 5, the variance is 5, and the standard deviation is $\sqrt{5}$. But if you now double the hours to $1 * 2 = 2$, where t = 2, Y is a Poisson distribution with $\lambda t = 5\,(2) = 10$ arrivals in 2 hours. The variance is also 10, and the standard deviation is the square root of 10, which is 3.16 arrivals in 2 hours.

REMEMBER

The mean and standard deviation are both in the same units of X (or Y); the variance is not. (The units of variance don't make sense most of the time since they are squared.)

WARNING

If you are operating in space instead of time, you have to be a little more careful when changing units. For example, if you are counting blemishes in 1 square yard of carpet and you want to double the length of the piece to 2 yards, you have to remember the square yardage now goes from $1 \times 1 = 1$ square yard to $2 \times 2 = 4$ square yards, and t = 4.

For example, suppose that you are counting blemishes in 1 square yard of carpet (X), and you know X is a Poisson distribution with $\lambda = 0.2$. You want to know the probability of getting two blemishes in 16 square yards (Y). You have already done the math and know that $4 \times 4 = 16$ square yards, and t = 16 here, not 4. There are 16 units now instead of just 1. So $\lambda t = 0.2(16) = 3.2$ and $P(Y = 2) = \dfrac{e^{-3.2}3.2^2}{2!} = 0.209$.

EXAMPLE

Q. If your carpet units are in square feet, and X is the number of blemishes per square foot in the carpet and has a Poisson distribution with $\lambda = 0.10$, what is λt for Y = the number of blemishes per square yard?

A. To go from square feet to square yards, you multiply by $3 \times 3 = 9$, so t = 9. That means λt for Y = the number of blemishes per square yard, which is $0.10 * 9 = 0.90$.

---

10 Suppose that you work at a helpline and you want to count the number of calls you get per hour, X. X is a Poisson distribution with $\lambda = 10$. What's the chance you get 20 calls in 2 hours?

11 Suppose that you are selling vegetables at a farmer's market. You count the number of people who stop by your booth in 1 hour, and you find it has a Poisson distribution with $\lambda = 15$. What is the mean and standard deviation of the number of visitors over the 4-hour period when the market is open?

---

12 Your neighbor has a yard sale and counts the number of people that arrive in 15 minutes and calls that number X. X is a Poisson distribution with $\lambda = 10$. What is the chance that at most 15 people arrive in 30 minutes?

Approximating the Poisson Distribution with the Normal Distribution

For values of λ from less than 1 up to 20, you can find probabilities for the Poisson distribution on a Poisson table. As the values of λ get larger, it gets harder to calculate them, and tables don't provide them. But the way the Poisson works, its pmf looks more and more like a normal distribution as λ gets larger. When λ is small, the Poisson distribution will be skewed to the right. As λ increases (generally considered to be > 20), the shape of the Poisson distribution becomes more symmetric and bell-shaped, resembling the normal distribution.

This means you can take the Poisson distribution and transform it to a standard normal distribution by subtracting the mean, λ, and dividing by the standard deviation, the square root of λ, to get Z. Then you can solve the problem like any other normal distribution problem using the Z table.

For example, suppose that the number of people in line for a very popular roller coaster has a Poisson distribution with $\lambda = 100$. You want to know the probability that X is at most 80. So you want $P(X \le 80)$, and because $\lambda > 20$, you use the normal approximation to solve this problem. You have:

$$P(X \le 80) = P\left(Z \le \frac{80 - \lambda}{\sqrt{\lambda}}\right) = P\left(Z \le \frac{80 - 100}{\sqrt{100}}\right) = P\left(Z \le \frac{-20}{10}\right) = P(Z \le -2) = 0.0228.$$

It's not very likely that fewer than 80 people are lined up for the roller coaster, so if you see that and you're at Disneyworld where the lines can be quite long, go for it!

To find $P(X > x)$ or $P(X \ge x)$, you rewrite the probability as a less-than-or-equal-to problem as you did when using the Poisson table, and take it from there. For example, $P(X > 80) = 1 - P(X \le 80) = 1 - 0.0228 = 0.9772$ from the previous example. To find $P(X \ge 80)$, you rewrite it as $1 - P(Z \le 79)$ and finish the problem from there to get

$$1 - P(Z \le 79) = 1 - P\left(Z \le \frac{79 - 100}{\sqrt{100}}\right) = 1 - P(Z \le -2.1) = 0.0179.$$

To find approximations for probabilities between two values, write them both as $P(X \le x)$ and subtract. For example, suppose that you want $P(90 < X \le 110)$ in the roller coaster example. You rewrite this probability as $P(X \le 110) - P(X \le 90)$. Using a number line to help make sure you have the right inequality signs and the correct numbers can really help. In this case, the number line looks like the number line shown in Figure 13-2.

FIGURE 13-2:
$P(90 < X \le 110) =$
$P(X \le 110) -$
$P(X \le 90)$
on a number
line.

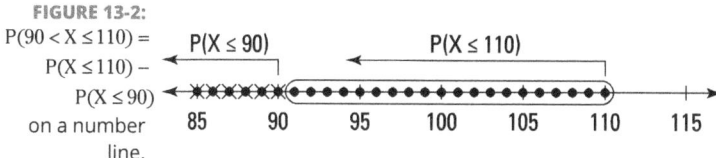

EXAMPLE

Q. You are counting the number of plane arrivals at the airport per hour. You figure X is a Poisson distribution with $\lambda = 40$. Find $P(X > 50)$.

A. $P(X > 50) = 1 = P(X \le 50) =$
$1 - P\left(Z \le \dfrac{50 - 40}{\sqrt{40}}\right) = 1 - P(Z \le 1.58) =$
$1 - 0.9429 = 0.0571$. So there is only about a 5.71 percent chance that more than 50 planes will arrive in 1 hour.

---

13 While at a dog park, Bentley counts the number of dogs that arrive at the dog park in 1 hour (X). The managers of the dog park know X is a Poisson distribution with $\lambda = 30$. What is the chance that at most 25 dogs will arrive in 1 hour?

14 The number of people who arrive in 1 hour for the keynote speech at a conference is a Poisson distribution with $\lambda = 100$. What's the chance that between 90 and 110 people (inclusive) will arrive in 1 hour?

---

15 Suppose that the number of typos in a 500-page book is a Poisson distribution with $\lambda = 25$. What's the chance that at least 30 typos exist in a 500-page book?

Solutions to Problems in Working with the Poisson (a Nonpoisonous) Distribution

(1) In this case, X is counting the number of arrivals over a fixed time period: 1 hour. The arrivals are independent, and no two can happen at the same time.

(2) $X = 0, 1, 2, 3, \ldots$.

(3) The binomial distribution fixes the number of trials, and the Poisson distribution fixes time or space. And binomial values go from $0, 1, 2, \ldots, n$, while Poisson values go from $0, 1, 2, 3, \ldots$.

(4) $P(X > 7) = 1 - P(X \le 7)$ where $\lambda = 5$. Using the Poisson table, you get $1 - 0.867 = 0.133$.

(5) $P(X < 6) = P(X \le 5)$ where $\lambda = 5$. Using the Poisson table, you get 0.616.

(6) $P(5 \le X \le 8) = P(X \le 8) - P(X \le 4)$ where $\lambda = 5$. Using the Poisson table, you get $0.932 - 0.440 = 0.492$. The following figure shows which values are included.

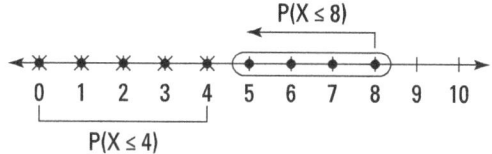

(7) If the mean of X is 4, and X has a Poisson distribution, the variance is also 4, and the standard deviation is the square root of 4, which is 2.

(8) You can see by the equation $\dfrac{e^{-16}16^{x}}{x!}$ that λ equals 16, so the variance, also λ, is 16, and the standard deviation is the square root of λ, which is 4.

(9) False. The mean and variance of the Poisson distribution are the same (λ), but the standard deviation is the square root of λ.

(10) X is the number of calls in 1 hour and is a Poisson distribution with $\lambda = 10$ calls. Y is the number of calls in 2 hours and is a Poisson process with $\lambda t = 10(2) = 20$. You want $P(Y = 20)$. Using the Poisson table, you have $P(Y = 20) = P(Y \le 20) - P(Y \le 19) = 0.559 - 0.470 = 0.089$.

(11) X is the number of visitors in 1 hour and is a Poisson distribution with $\lambda = 15$ people. Y is the number of visitors in 4 hours and is a Poisson process with $\lambda t = 15(4) = 60$ people. The mean is $\lambda t = 60$, and the standard deviation is the square root of λt, which is $\sqrt{60} = 7.75$.

(12) X is the number of arrivals in 15 minutes and is a Poisson distribution with $\lambda = 10$ people. The number of arrivals in 30 minutes is a Poisson process with $\lambda t = 10(2) = 20$. You want $P(X \le 15) = 0.157$ using the Poisson table.

(13) X is the number of dog arrivals in 1 hour and is a Poisson distribution with $\lambda > 30$. You want $P(X \le 25)$. Using the normal approximation, you have.
$$P(X \le 25) = P\left(Z \le \frac{25 - 30}{\sqrt{30}}\right) = P(Z \le -0.91) = 0.1814.$$

(14) X is a Poisson distribution with $\lambda = 100$. Using the normal approximation, you have
$$P(90 \le X \le 110) = P\left(\frac{90 - 100}{\sqrt{100}} \le Z \le \frac{110 - 100}{\sqrt{100}}\right) = P(-1 \le Z \le 1) = 0.8413 - 0.1587 = 0.6826.$$

(15) X is a Poisson distribution with $\lambda = 25$. You want $P(X \ge 30)$. This is approximately equal to
$$P(X \ge 30) = P\left(Z \ge \frac{30 - 25}{\sqrt{25}}\right) = P(Z \ge 1) = 1 - 0.8413 = 0.1587.$$

Chapter **14**

Covering All Angles of the Geometric Distribution

The geometric distribution may remind you of one of our old friends from Chapter 8, the binomial distribution. (If you've not covered that section yet, don't worry; we start from scratch in this chapter.) While the binomial distribution counts the number of times a certain event occurs in n fixed trials, the geometric distribution counts the number of trials until a certain event occurs. For example, instead of counting the number of heads in ten coin flips, as the binomial does, the geometric counts the number of flips needed to get ten heads.

In this chapter, you unlock the geometric distribution's characteristics; examine its probabilities; probe its mean, variance, and standard deviation; and work some problems. Sound like a plan?

Characteristics of the Geometric Distribution

The geometric distribution is based on the geometric series from math. It is a sum of the values of a fraction between 0 and 1 taken to higher and higher powers, starting with the fraction to the zero power, which is 1. For example, one geometric series sums the values of ½ taken to higher and higher powers: $(1/2)^0 + (1/2)^1 + (1/2)^2 + (1/2)^3 + \ldots$, and so on, into infinity.

The sum of the values of this series is $\dfrac{1}{1-r}$, where r is that fraction between 0 and 1. In our case, the value of r is ½, so the sum of the values in the infinite series is $\dfrac{1}{1-1/2}$ = 2. Now, if you remove the first term of the series — the value of ½ to the zero power, which is 1 (because it can't take 0 trials to get anywhere) — you get $(1/2)^1 + (1/2)^2 + (1/2)^3 + \ldots$, which equals $2 - 1 = 1$, and this is a fitting probability distribution for a random variable. This distribution is called the *geometric distribution*, in honor of this infinite series (modified).

TIP

By the way, if you haven't covered the math to fully understand this explanation, that's okay. The important aspects to recognize are that the geometric distribution models the number of trials until the first success occurs of a certain event, such as getting a head, making a basket, or winning money on a pull tab, provided certain characteristics hold.

The important characteristics that identify the geometric distribution are the following:

>> The problem includes a sequence of independent trials of some random process. (The number of trials is unknown.)

>> Each outcome has one of two options: success (when the event of interest occurred) or failure (when the event of interest did not occur).

>> The probability of success, p, is the same for each trial and does not change. This means $1 - p$ is the chance for failure on a trial.

>> X counts the number of trials up to and including the first success (which means the number of failures is X – 1).

For example, let's look at the action of flipping a coin until it lands on heads. You have a sequence of trials of the random flips; each outcome is a success (head) or failure (tail); the probability of success, assuming the coin is fair, is ½, and does not change from flip to flip. The probability of failure, assuming the coin is fair, is $1 - ½ = ½$, and does not change from flip to flip. X counts the number of flips up to and including the first success (a head). So, say you flipped the following sequence: TTTTTH. That means X is 6 because you got your first head on the sixth flip. (It also means the number of tails is X – 1 = 5.)

EXAMPLE

Q. Tell whether the following statement is true or false: "One of the characteristics of the geometric distribution is that it counts the number of successes in n trials."

A. False. The geometric distribution counts the number of trials up to and including the first success.

1. Tell whether the following statement is true or false: "The series $1 + 2 + 4 + 8 + \ldots$ is a geometric series."

2. Tell whether the following statement is true or false: "You have only two possible outcomes on any trial of a geometric distribution."

3. Which of the following situations has a geometric distribution? Choose all that apply.

   a. The number of free throws shot by various players in a third-grade basketball game until a free throw is made.

   b. The number of men who get chosen for a committee before the first woman is chosen. (Assume ten men and ten women to start.)

   c. The number of people you have to randomly choose before you get someone who has won a lottery.

   d. The number of flips of a coin you have to make until you get two heads.

   e. The number of heads you get when you flip a coin for 10 seconds.

Finding Probabilities for the Geometric Distribution

In this section, you find out how to calculate probabilities for a geometric distribution. We start with the probability mass function (pmf) for the geometric distribution, which gives you the values of X along with the probabilities of X in general form.

The formula for the probability mass function for the geometric distribution is $P(x) = (1-p)^{x-1}p$, for $x = 1, 2, 3, \ldots$, where the number of trials up to and including the first success is x, the number of failures before the first success is $x - 1$, the probability of success is p, and the probability of failure is $1 - p$.

For example, suppose that you are flipping a coin and you get a head on the eighth try. You might think that getting seven tails in a row is unbelievable — what are the chances of that? Well, you can find out by taking $p = \frac{1}{2}$, $1 - p = \frac{1}{2}$, and $x = 8$. You get the following: $P(8) = (1-1/2)^{8-1}1/2 = (1/2)^8 = 0.0039$. That's 39 out of 10,000! Your sequence would look like this: TTTTTTTH.

TIP

The geometric series closes in on you and gets smaller and smaller pretty quickly by multiplying all those failures together over and over.

EXAMPLE

Q. Suppose that you roll a fair six-sided die and are looking to get a 1. What is the chance it happens (for the first time) on your third try?

A. Here, success is getting a 1; failure is not getting a 1 (meaning 2, 3, 4, 5, or 6). So, for example, your sequence could have been 2, 6, 1, or 3, 3, 1. The value of p is $\frac{1}{6}$ = the probability of getting a 1, and the value of $1 - p$ is $\frac{5}{6}$, the probability of not getting a 1. X is 3 because you're getting your first one on the third try. So the probability of this happening is

$$P(3) = \left(1-\frac{1}{6}\right)^2 * \frac{1}{6} = \left(\frac{5}{6}\right)^2 * \frac{1}{6} = 0.1157.$$

Suppose that you buy five random scratch-off lottery tickets, each of the same kind.

a. If the chance of winning any prize is $\frac{1}{10}$, what is the chance that you win something on the first try?

b. What's the chance that you win something on the second try (but not the first try)?

c. Fill out the following table of probabilities for the first through fifth tries. Do you see a pattern?

Try	Probability
1	
2	
3	
4	
5	

 Suppose that you are flipping through channels on the TV, and there is a $\frac{1}{10}$ chance you'll land on a show that you like. What's the chance you flip through five shows before you find one that you like on the sixth try?

6 The chance of an overdue book being in the return pile at the library is 0.6. When you are putting away a random selection of books, what is the chance that you find your first overdue book on the second try?

Probing the Mean, Variance, and Expected Value of the Geometric Distribution

The mean of X when X is a geometric distribution answers the "how long?" question. *How long* do you expect to keep doing some action until you get your first success? The answer is $\frac{1}{p}$. That's the formula for the expected value (the mean) of the geometric distribution when p is your probability of success. (Too bad the math to derive this formula isn't so easy, but the good news is that we get to use the results!) For example, if you are flipping a fair coin until the coin lands on tails, you expect in the long run to flip the coin $\frac{1}{(\frac{1}{2})} = 2$ times until you get your first heads. Now, does this actually happen every time in real life? No. That's why Las Vegas is full of beautiful casino compounds!

Similarly, how long do you expect to keep shooting craps until a 7 comes up? Well, you have to figure out p, the chance of getting a 7 in the first place (with two dice), and flip it over. When rolling two dice, and looking at the combinations of 36 pairs of results (6 × 6 on each die), you get the following table of possible values for the sum of the dice, which is of interest in the game of craps:

Sum of 2 Dice	Probability
2	$P(1, 1) = \frac{1}{36}$
3	$P(1, 2) + P(2, 1) = \frac{2}{36}$
4	$P(1, 3) + P(2, 2) + P(3, 1) = \frac{3}{36}$
5	$P(1, 4) + P(2, 3) + P(3, 2) + P(4, 1) = \frac{4}{36}$
6	$P(1, 5) + P(2, 4) + P(3, 3) + P(4, 2) + P(5, 1) = \frac{5}{36}$
7	$P(1, 6) + P(2, 5) + P(3, 4) + P(4, 3) + P(5, 2) + P(6, 1) = \frac{6}{36}$
8	$P(2, 6) + P(3, 5) + P(4, 4) + P(5, 3) + P(6, 2) = \frac{5}{36}$
9	$P(3, 6) + P(4, 5) + P(5, 4) + P(6, 3) = \frac{4}{36}$
10	$P(4, 6) + P(5, 5) + P(6, 4) = \frac{3}{36}$
11	$P(5, 6) + P(6, 5) = \frac{2}{36}$
12	$P(6, 6) = \frac{1}{36}$

Notice that the probabilities are symmetric, starting at $\frac{1}{36}$ at 2, going up to $\frac{6}{36}$ at 7, then back down to $\frac{1}{36}$ at 12. The peak is $\frac{6}{36}$, which happens when the sum is, you guessed it, 7. So if you want to know how long it takes to get a 7 in craps (assuming you believe rolls are independent, which I do), then your answer is $\frac{1}{(\frac{6}{36})} = 6$. So, about every six rolls in the long run. Now that's the expected value — what actually happens in real life is what keeps the crowds a-comin' to the tables every night in Vegas.

The variance of X when X is geometric is $\frac{1-p}{p^2}$, where p is the probability of success. In the case of craps, the variance is $\frac{1-\frac{6}{36}}{\left(\frac{6}{36}\right)^2} = 30$, and the standard deviation is the square root of 30, which is 5.48.

REMEMBER

If the probability of failure $(1-p)$ is small, then the probability of success (squared in the denominator) is large, so the overall variance is small. This makes sense because you expect your successes rather quickly. If the probability of failure is large, the probability of success is small, so the overall variance is large. There are more places to fit the first success in that case.

The standard deviation of X when X is geometric is the square root of the variance, which is $\sqrt{\frac{1-p}{p^2}}$. It is in the same units of X.

EXAMPLE

Q. What is the variance and standard deviation of the number of trials until the first success when you are flipping a fair coin and looking for your first heads?

A. Here $p = \frac{1}{2}$, is the probability of a success (heads), so the variance is $\frac{1-p}{p^2} = \frac{(1-\frac{1}{2})}{(\frac{1}{2})^2} = 2$, and the standard deviation is the square root of 2, which is 1.414.

 7 What is the expected number of rolls of a (fair) die until you get a 6?

 8 What is the standard deviation of the number of rolls of a die until you get a 1?

9 What is the expected number of rolls of a die until you get a 1 or a 6?

10 What is the variance of the number of rolls of a die until you get a 1 or a 6?

11 What is the standard deviation of the number of rolls of a die until you get a 1 or a 6?

12 Tell whether the following statement is true or false: "If the probability of failure $(1-p)$ is small, then the overall variance is small."

13 Tell whether the following statement is true or false: "If the probability of failure is large, the probability of success is small, so the overall variance is large."

14 Tell whether the following statement is true or false: "If the probability of success is large, the mean will also be large."

15 The chance of an overdue book being in the return pile at the library is 0.6. Let X be the number of books you go through until you find your first overdue book. What is the mean, variance, and standard deviation of X?

Solutions to Problems in Covering All Angles of the Geometric Distribution

(1) False. The geometric series is based on a fraction between 0 and 1 taken to higher and higher powers. In this example, the number 2 is being taken to higher and higher powers, and 2 is not a fraction between 0 and 1.

(2) True. Each outcome has one of two options: success (where the event of interest occurred) or failure (where the event of interest did not occur).

(3) a. This is not a geometric distribution. The value of p is different for different players.

b. This is not a geometric distribution because n can only be 1 to 10. Geometric distributions are 1, 2, 3, and so on.

c. This may be considered a geometric distribution. Because there are so many people who play the lottery, you might as well call it infinite.

d. This is not a geometric distribution. You are only supposed to go until you get the first head, not two heads.

e. This is not a geometric distribution. Time is not fixed in a geometric distribution.

(4) a. First try: $\frac{1}{10}$.

b. Second try (but not first try): $0.9*0.1 = 0.09$

c. The pattern is teach time you add a trial, you multiply by $\frac{9}{10}$, so each value is $\frac{9}{10}$ times the previous one:

Try	Probability
1	$\frac{1}{10}$
2	$\frac{9}{10} * \frac{1}{10}$
3	$\left(\frac{9}{10}\right)^2 * \frac{1}{10}$
4	$\left(\frac{9}{10}\right)^3 * \frac{1}{10}$
5	$\left(\frac{9}{10}\right)^4 * \frac{1}{10}$

(5) $\left(\frac{9}{10}\right)^5 * \frac{1}{10} = 0.059$.

(6) $0.4*0.6 = 0.24$.

(7) $p = \frac{1}{6}$, so the mean $\mu_x = \dfrac{1}{\left(\frac{1}{6}\right)} = 6$ rolls.

(8) $p = \frac{1}{6}$, so the standard deviation $\sigma_x = \sqrt{\dfrac{1-\frac{1}{6}}{\left(\frac{1}{6}\right)^2}} = 5.48$.

(9) $p = \frac{2}{6} = \frac{1}{3}$, since it could be a 1 or a 6. The mean is $\mu_x = \dfrac{1}{\left(\frac{1}{3}\right)} = 3$ rolls.

(10) $p = \frac{2}{6}$, so $\sigma_x^2 = \frac{1 - \frac{2}{6}}{\left(\frac{2}{6}\right)^2} = 6$.

(11) $p = \frac{2}{6}$, so $\sigma_x = \sqrt{\frac{1 - \frac{2}{6}}{\left(\frac{2}{6}\right)^2}} = 2.45$.

(12) True. The probability of failure is in the numerator of the variance.

(13) True. You have a large number divided by the square of a small number in the variance, which makes the overall value larger.

(14) False. If p is large, then $\frac{1}{p}$ = the mean number of trials until the first success, is small. In other words, p being large means a success is coming up soon, so there isn't room for a lot of variation.

(15) The mean is $\frac{1}{p} = \frac{1}{0.6} = 1.67$. The variance is $\frac{(1-p)}{p^2} = \frac{(0.4)}{(0.6)^2} = 1.11$, and the standard deviation of X is the square root of 1.11, which is 1.05.

Chapter **15**

Making a Positive Out of the Negative Binomial Distribution

The negative binomial distribution is a more general version of the geometric distribution discussed in Chapter 14. Instead of counting the total number of trials until the first success, you count the number of trials until the kth success, where k is any number from 1, 2, 3, and so on. For example, the geometric distribution counts the number of flips of a coin until the first head appears, but the negative binomial distribution counts the number of flips of a coin until the third heads (or fourth heads or tenth heads, and so on).

Perhaps it is called a negative binomial because, in a way, it's the opposite of a regular binomial distribution (see Chapter 8). The binomial fixes the number of trials and counts the number of successes within those trials; for example, you flip a coin 25 times (the fixed part) and get $X = 10$ heads (the random part). Whereas the negative binomial fixes the number of successes and counts the number of trials up to and including the kth success; for example, you flip a coin $X = 25$ times (the random part) until the tenth heads (the fixed part). They sound similar but are very different.

In this chapter, you examine the characteristics of the negative binomial distribution, find probabilities for it, as well as the mean, variance, and standard deviation.

Checking to See When You Have a Negative Binomial Distribution

The following list outlines the characteristics of a negative binomial distribution, X:

>> You observe a sequence of independent trials from some random process, like flipping a coin, for example.

>> You classify the outcomes of each trial into two groups: success and failure.

>> The probability of success is the same for each trial; let p be the probability of success and $1-p$ be the probability of failure.

>> X counts the total number of trials up to and including the kth success; there are $x-k$ total failures since x is the total number of trials. For example, x might be ten flips until the fourth success. Here $x=10$, $k=4$ successes, and $x-k=10-4=6$ failures.

TIP

Note that some textbooks use the letter q to denote the probability of failure rather than $1-p$.

EXAMPLE

Q. Tell whether the following statement is true or false: "The fixed part of the negative binomial distribution is the number of successes, while the random part is the number of trials needed to get those k successes (counting both successes and failures)."

A. True. This is the definition of the negative binomial distribution.

 Suppose that you count the number of students in a classroom of 20 students who are chosen up through and including the third female student. Do you have a negative binomial distribution?

2 Tell whether the following statement is true or false: "The negative binomial distribution is the same as the geometric distribution when k (the number of successes) equals 1."

3 Tell whether the following statement is true or false: "X counts the total number of failures until the kth success in a negative binomial distribution."

4 Tell whether the following statement is true or false: "In a binomial distribution, the number of trials is known (fixed), but in a negative binomial distribution, the number of trials is unknown (random)."

5 Tell whether the following statement is true or false: "If you have unlimited M&M's candies, and you randomly select from them until you get six blue M&M's, you have a negative binomial distribution."

Noting Probabilities for a Negative Binomial

The formula for finding the probability for X where X has a negative binomial distribution, is given by the following: $P(X = x) = \binom{x-1}{k-1}p^k(1-p)^{x-k}$, where the following conditions hold:

» x is the number of trials up through and including the kth success. The lowest number x can be is k, and that happens when you get k successes all in a row.

» $x = k, k + 1, k + 2, ...,$ to infinity.

» k is at least 1.

» p is the probability of success, and $(1 - p)$ is the probability of failure.

» k is the number of successes desired.

» There are $\binom{x-1}{k-1}$ ways to fill in the rest of the spots with successes, after you know that the last spot must be a success and the process stops.

» X is the number of trials it takes to achieve the k successes, including the kth success.

For example, suppose that you want the probability of rolling a die until three 6s come up. What is the chance that it takes three rolls? In that case, you'd have to get 6, 6, 6 right away. The probability turns out to be: $P(X = 3) = \binom{3-1}{3-1}(\frac{1}{6})^3(1-\frac{1}{6})^0 = (\frac{1}{6})^3 = 0.0046$.

Now, what if you up the number of trials to four? That is, find $P(X = 4)$. Here $X = 4$ and $k = 3$, so you have $P(X = 4) = \binom{4-1}{3-1}(\frac{1}{6})^3(1-\frac{1}{6})^1 = 3(\frac{1}{6})^3(\frac{5}{6}) = 0.0116$. This probability is higher because you have more arrangements that work, and you have to end with the third 6:

    non-6, 6, 6, 6

    6, non-6, 6, 6

    6, 6, non-six, 6

As you increase the number of trials up to and including the third 6, the probabilities will increase to a point, then they will start to back down near the mean, as the number of non-6s begins to take over, and larger amounts of them are harder to get.

EXAMPLE

Q. Suppose that the chance of a factory creating a defective part is 0.10. What is the chance of randomly sampling ten parts until you get two defectives (where the second defective part was chosen tenth)?

A. You have p = 0.10, x = 10 trials, k = 2 (defectives = successes), x − k = 8 failures (non-defectives), so
$$P(X = 10) = \binom{10-1}{2-1} \cdot 1^2(0.9)^8 =$$
$$\binom{9}{1} \cdot 1^2(0.9)^8 = 0.0387.$$

6   Tell whether the following statement is true or false: "If k = 1, you have a binomial distribution."

7   Why do the values of x start with k?

8   Suppose that the chance of winning a scratch-off lottery ticket is one in ten. What is the chance of randomly sampling ten tickets until you get not one, but two winners?

9   Baron guesses the outcome of a rolled die until he gets three of them right. What's the chance that he can do this in exactly ten tries?

10   You're playing a card game in which you're supposed to guess the suit of a chosen card (diamond, heart, club, spade). What's the chance you correctly guess the suit of five cards out of ten tries, and the last try is correct?

Figuring the Mean, Variance, and Standard Deviation of the Negative Binomial

The mean number of trials needed to obtain k successes (where the last trial is a success) is $\mu = \dfrac{k}{p}$. The probability of success, p, is the determining factor here because k is fixed. A higher value of p means a smaller mean number of trials is needed to obtain k successes, and a lower value of p means a larger mean number of trials is needed to obtain k successes.

REMEMBER

$\dfrac{k}{p}$ is just k times the formula for the mean of the geometric distribution, 1/p (see Chapter 14), since the geometric has only one success at the end, meaning k = 1.

Suppose that a certain scratch-off lottery ticket has a probability of winning of ¼. What is the mean number of scratch-off tickets you have to go through on average to win two times and then stop on the second win? Here, k = 2 and p = ¼, so μ would be $\dfrac{k}{p} = \dfrac{2}{\frac{1}{4}} = 8$ tickets. How many do you go through on average to win three times, then stop on the third win? Here, k = 3 and p = ¼, so μ would be $\dfrac{k}{p} = \dfrac{3}{\frac{1}{4}} = 12$, and so on.

The average isn't the exact number of tickets you'll have to go through until you win k times. It's an average; sometimes it's less, sometimes it's more. It depends on the standard deviation as well, which is coming up.

The formula for the variance of the negative binomial distribution is $\sigma^2 = \dfrac{k(1-p)}{p^2}$; the standard deviation is $\sigma = \sqrt{\dfrac{k(1-p)}{p^2}}$.

For the scratch-off lottery ticket example, the probability of winning is ¼. What is the variance when you win two times and then stop? (The mean is 8.) You get $\sigma^2 = \dfrac{k(1-p)}{p^2} = \dfrac{2(1-\frac{1}{4})}{(\frac{1}{4})^2} = 24$. The standard deviation is the square root of 24, which is 4.90.

What if you win a total of three times and then stop on the third win? What is the variance and standard deviation of the total number of tries? Here, p = ¼ and k = 3. You have $\sigma^2 = \dfrac{k(1-p)}{p^2} = \dfrac{3(1-\frac{1}{4})}{(\frac{1}{4})^2} = 36$. (The mean is 12.) The standard deviation is the square root of 36, which is 6.

Q. Suppose that the chance of a factory creating a defective part is 0.10. What is the mean number of trials of randomly sampled parts you would have to do until you get two defectives?

A. You have p = 0.10 and k = 2 (defectives = successes). The mean is $\mu = \dfrac{k}{p} = \dfrac{2}{0.10} = 20$ trials.

---

11 What is the average number of card suits you have to guess in a row until you guess the right suit for the fifth time and stop there?

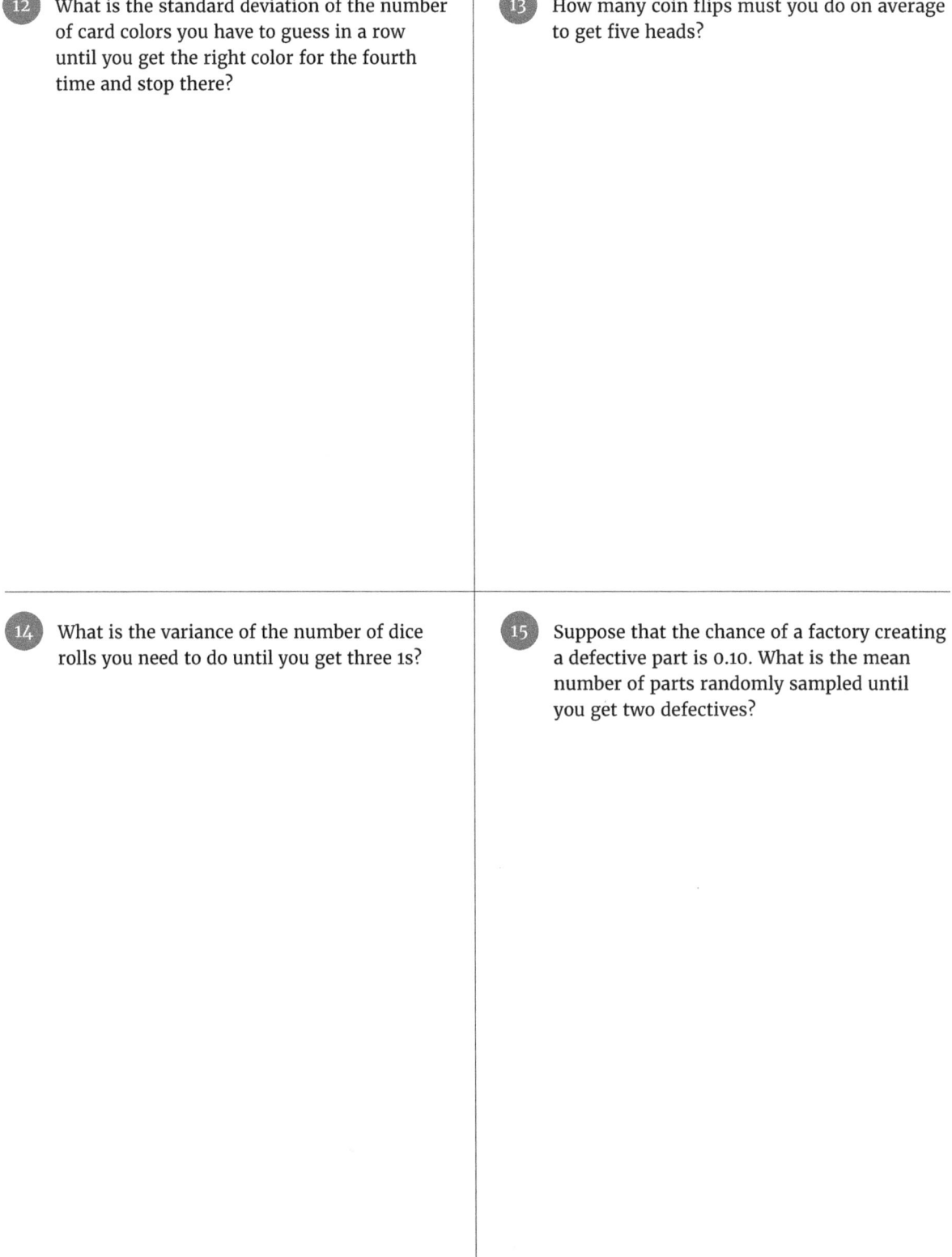

12 What is the standard deviation of the number of card colors you have to guess in a row until you get the right color for the fourth time and stop there?

13 How many coin flips must you do on average to get five heads?

14 What is the variance of the number of dice rolls you need to do until you get three 1s?

15 Suppose that the chance of a factory creating a defective part is 0.10. What is the mean number of parts randomly sampled until you get two defectives?

Solutions to Problems in Making a Positive Out of the Negative Binomial Distribution

(1) No. You do not have a negative binomial random variable. The value of p is not the same for each trial. You are sampling without replacement.

(2) True. The negative binomial distribution is the same as the geometric distribution when k (the number of successes) equals 1.

(3) False. X counts the total number of trials until the kth success in a negative binomial distribution, not the number of failures.

(4) True. In a binomial distribution, the number of trials is known (fixed), but in a negative binomial distribution, the number of trials is unknown (random).

(5) True. If you have unlimited M&M's, and you randomly select from them until you get six blue M&M's, you have a negative binomial distribution so long as X is counting the total number of trials up through and including the sixth blue M&M.

(6) False. If k = 1, you have a geometric distribution.

(7) The possible values of x start with k = number of successes because x counts the number of trials up through and including the kth success, and the minimum number of trials in that case is k (and occurs when you get all the successes).

(8) The formula is $P(X = x) = \binom{x-1}{k-1} p^k (1-p)^{x-k}$. In this problem, $p = \frac{1}{10}$, k = 2, and x = 10, so

$$P(X = 10) = \binom{10-1}{2-1} \cdot 1^2 (0.9)^8 = \binom{9}{1} \cdot 1^2 (0.9)^8 = 0.0387.$$

(9) The formula is $P(X = x) = \binom{x-1}{k-1} p^k (1-p)^{x-k}$. In this problem, $p = \frac{1}{6}$, k = 3, and x = 10, so

$$P(X = 10) = \binom{10-1}{3-1}(\tfrac{1}{6})^3(\tfrac{5}{6})^7 = \binom{9}{2}(\tfrac{1}{6})^3(\tfrac{5}{6})^7 = 36(0.0013) = 0.0465.$$

(10) The formula is $P(X = x) = \binom{x-1}{k-1} p^k (1-p)^{x-k}$. In this problem, $p = \frac{1}{4}$ (since there are four suits), k = 5, and x = 10, so $P(X = 10) = \binom{10-1}{5-1}(\tfrac{1}{4})^5(\tfrac{3}{4})^5 = \binom{9}{4}(\tfrac{1}{4})^5(\tfrac{3}{4})^5 = 126(0.0013) = 0.0292.$

(11) In this problem, $p = \frac{1}{4}$ and k = 5, so the mean is $\mu = \frac{k}{p} = \frac{5}{\frac{1}{4}} = 20.$

(12) In this problem, $p = \frac{1}{2}$ since there are two colors, and k = 4, so the standard deviation is

$$\sigma = \sqrt{\frac{k(1-p)}{p^2}} = \sqrt{\frac{4(1-0.5)}{0.5^2}} = 2.83.$$

(13) In this problem, $p = \frac{1}{2}$ and k = 5, so the mean is $\frac{k}{p} = \frac{5}{\frac{1}{2}} = 10.$

(14) In this problem, $p = \frac{1}{6}$ and k = 3, so the variance is $\sigma^2 = \frac{3(1-\frac{1}{6})}{(\frac{1}{6})^2} = 90$. (The standard deviation is 9.49.)

(15) In this problem, p = 0.10 and k = 2, so ts $\mu = \frac{k}{p} = \frac{2}{0.10} = 20.$

Chapter **16**

Not Getting Hyper about the Hypergeometric Distribution

The hypergeometric distribution is different than the other distributions covered in this book. In a hypergeometric situation, you pull a sample and divide it into two groups: those you are interested in and those you aren't. It's based on combinations. You do it all at once, rather than in a series of trials. So you're doing it without replacement.

In this chapter, you explore the characteristics of the hypergeometric distribution, find probabilities for it, and measure the mean, variance, and standard deviation. The big picture involves breaking the population into two subgroups: the subgroup that has the characteristic you are interested in and the subgroup that doesn't have the characteristic you are interested in, and then taking a sample from each subgroup.

Identifying the Hypergeometric Distribution

A random variable X must meet the following conditions to have a hypergeometric distribution:

>> You sample without replacement from a population of N fixed total individuals.

>> Every individual in the population has an equal chance of being sampled.

>> You can classify every individual in the population into one of two subgroups or subpopulations:

- *The marked subpopulation:* Individuals who have the characteristic you are interested in (for example, rent their home, own a pet (or pets), or like to eat pizza).

- *The unmarked subpopulation:* Everyone else in the population who isn't in the marked subpopulation (for example, don't rent their own home, don't own any pets, don't like to eat pizza).

>> The total number of marked individuals in the population is fixed and known as M.

>> X counts the number of marked individuals in the sample. You find probabilities, the mean, the variance, and the standard deviation of X.

For example, suppose that you are in a club of 20 students, and among the members are 14 women and 6 men. The group randomly selects five people to be on a committee, and it turns out that the people selected are all men. You ask, "What are the odds of that happening?" This is a situation for the hypergeometric distribution: you have N = 20 people in the population broken into two groups, M = 6 individuals are men (since that is who you are counting), everyone has an equal chance of being selected, and you are sampling them without replacement (once you choose a person, they can't be chosen again). This last fact is what separates the hypergeometric distribution from the binomial distribution (see Chapter 8).

EXAMPLE

Q. Tell whether the following statement is true or false: "In the hypergeometric distribution, you sample without replacement, and with the binomial, you sample with replacement."

A. True. These are part of the conditions of each of the distributions.

1. Tell whether the following statement is true or false: "In the hypergeometric distribution, $M < N$."

2. Tell whether the following statement is true or false: "The total number of individuals in the unmarked population must be $N - M$."

3. Tell whether the following statement is true or false: "Sampling one individual at a time with replacement is the same as reaching into the population and choosing your sample all at once without replacement."

4. At the end of the fiscal year party, Bob gives out three prizes at random to his staff of ten people, six of whom are women and four of whom are men. A person may win more than one prize. X counts the number of women who get a prize. Is this a hypergeometric distribution?

5 At the end of the fiscal year party, Bob gives out prizes at random to his staff of ten people, six of whom are women and four of whom are men. A person may not win more than one prize. X counts the number of women who get a prize. Is this a hypergeometric distribution?

6 At the end of the fiscal year party, Bob gives out prizes at random to his staff of ten people, six of whom are women and four of whom are men. X counts the number of prizes given away until a woman wins a prize. Is X a hypergeometric distribution?

Producing Probabilities for the Hypergeometric Distribution

The hypergeometric distribution uses combinations to find probabilities (see Chapter 5 for more on combinations). A *combination* is a formula used to figure out the number of ways to select a certain number of items from a group without replacement, where the order of the items does not matter. In this section, you discover how combinations become a part of the hypergeometric probability mass function (pmf). You also read about the domain for X (which is a list of its possible values) and walk through step-by-step calculations of probabilities.

Setting up the hypergeometric probability mass function

The probability mass function (pmf) of the hypergeometric distribution is:

$$P(X = x) = \frac{\binom{M}{x}\binom{N-M}{n-x}}{\binom{N}{n}}$$

where: $\max(0, M + n - N) \le x \le \min(M, n)$.

Following is the breakdown of all the variables in this formula (we dissect the range for the possible values of x in the next section):

» N is the number of total individuals in the population.

» n is the number of total individuals in the sample.

» M is the number of marked individuals in the population.

» x is the number of marked individuals in the sample.

» N – M is the number of unmarked individuals in the population.

» n – x is the number of unmarked individuals in the sample.

TIP

Note that the numbers on top in the numerator, M, and N – M sum to the number at the top of the denominator, N. That's because the population (of size N) is divided into the two sub-groups: the marked individuals (of which there are M) and the unmarked individuals (of which there are N – M). Notice also that the numbers on the bottom in the numerator, x and n – x, sum to the number at the bottom of the denominator, n. That's because the sample (of size n) is divided into two subgroups: the marked individuals in the sample (x) and the unmarked individuals in the sample (n – x).

To get the probability that P(X = x), you divide the population into the M marked cards and the N – M unmarked cards. In the numerator, you count the number of ways to choose x out of the M marked cards and (n – x) out of the N – M unmarked cards. In the denominator, you put them all together and count the number of ways to simply choose n cards out of the N cards in the whole population.

For example, Derrick has taken all 13 clubs from a 52-card deck as his population. He draws three cards from that group. What is the chance that two of them are face cards? (A face card is a jack, queen, or king.) Derrick has 13 cards in the population, so N = 13, and 3 are marked as face cards (so M = 3). The rest are unmarked as non-face cards (so N – M = 13 – 3). He samples a total of n = 3 cards, two from the marked cards (face cards) and one from the unmarked cards (non-face cards). He uses the following formula to find his probability:

$$P(X = x) = \frac{\binom{M}{x}\binom{N-M}{n-x}}{\binom{N}{n}} \rightarrow P(X = 3) = \frac{\binom{3}{2}\binom{13-3}{3-2}}{\binom{13}{3}}$$

$$= \frac{\binom{3}{2}\binom{10}{1}}{\binom{13}{3}} = \frac{3(10)}{(13)\,(12)\,(11)\,(10!)\,/\,(3)\,(2)\,(1)\,(10!)} = \frac{30}{286} = 0.1049.$$

Understanding the boundary conditions (domain) of X

If you want to write out the entire probability distribution for X, you need to know the possible values for X, which is also known as the *domain of X*.

The values in the domain for X in the hypergeometric distribution take on the following: $\max(0, M + n - N) \le x \le \min(M, n)$. That's a mouthful for sure. But let's break it down so that it's easier to understand.

X lies between two boundaries at the same time:

» x has to be less than or equal to the minimum of the two values, M and n.

» x has to be greater than or equal to the maximum of the two values 0, and $M + n - N$.

The boundary conditions can be explained by looking at the boundaries on the number of marked and unmarked individuals in the sample:

» x is the number of marked individuals in the sample. Therefore, x has to be less than or equal to the sample size, n, but it also has to be less than or equal to the total marked individuals in the population, M. So x must be less than or equal to $\min(M, n)$.

» $n - x$ is the number of unmarked individuals in the sample. Therefore, $n - x$ has to be less than or equal to the number of unmarked individuals in the population, which is N – M. This means $n - x \le N - M$, or $x \ge M + n - N$. It also has to be greater than or equal to 0, so in the end, x must be greater than or equal to the $\max(0, M + n - N)$.

Q. An office staff is made up of seven individuals, four of whom are men and three of whom are women. A committee of two is chosen at random. What is the chance that both individuals chosen are women?

A. Marked = women and M = 3, Unmarked = men and N – M = 7 – 3 = 4. Here, x = 2 so $n - x = 2 - 2 = 0$. N = 7 and n = 2 (the sample size). You want

$$P(X = x) = \frac{\binom{M}{x}\binom{N-M}{n-x}}{\binom{N}{n}} = \frac{\binom{3}{2}\binom{7-3}{2-2}}{\binom{7}{2}} =$$

$$\frac{\binom{3}{2}\binom{4}{0}}{\binom{7}{2}} = \frac{3(1)}{\left(\frac{7!}{2!5!}\right)} = 3/21 = 0.1429.$$

7　Suppose that you are throwing a party for 11 people, and you are giving away three prizes. Five people at the party live in your neighborhood, and six people at the party live outside your neighborhood. What's the chance that two of the prizes go to people in your neighborhood?

8　Suppose that you are throwing a party for 11 people, and you are giving away three prizes. Five people at the party live in your neighborhood, and six people at the party life outside your neighborhood. Let X = the number of prizes given to people in your neighborhood. Find the pmf of X, including the boundaries for the domain of X.

9　A bowl of Halloween candy contains ten fun-sized Snickers candy bars and eight fun-sized Milky Way candy bars. A child reaches into the bowl and takes a random handful of four candy bars. What are the boundary conditions for X, the number of Milky Way bars selected?

10　A bowl of Halloween candy contains ten fun-sized Snickers candy bars and eight fun-sized Milky Way candy bars. A child reaches into the bowl and takes a random handful of four candy bars. What is the chance the child selects all Milky Way bars?

11 Shelbi sells two types of bread: sourdough and rye. Today she made five loaves of sourdough and four loaves of rye. What is the chance that someone comes in and randomly buys three loaves of each?

Counting on the Expected Value, Variance, and Standard Deviation of the Hypergeometric

The mean (or expected value) of the hypergeometric distribution is $n\dfrac{M}{N}$, which is the sample size times the proportion of marked individuals in the population. In other words, it represents the expected number of marked individuals in the sample. For example, if you have a bowl of candy containing ten Snickers bars and eight Milky Way bars and you reach in and pull out a sample of four candy bars, the expected number of Milky Way bars (X) is $4*\dfrac{8}{13}=\dfrac{32}{18}=1.78$. The expected number of Snickers bars (if that's what you are marking, or counting) is $4*\dfrac{10}{18}=\dfrac{40}{18}=2.22$.

The variance of the hypergeometric is not an intuitive formula, but it does involve N, M, and n. Here is the formula for the variance:

$$\dfrac{nM}{N^2(N-1)}(N-M)(N-n).$$

For example, the variance of X, where X is the number of Snickers bars in the sample of four candy bars is $\dfrac{nM}{N^2(N-1)}(N-M)(N-n)=\dfrac{(4)(10)}{18^2(18-1)}(18-10)(18-4)=0.8133$.

The standard deviation of the hypergeometric distribution is the square root of the variance, which is: $\sqrt{\dfrac{nM}{N^2(N-1)}(N-M)(N-n)}$. In this case, the standard deviation is the square root of 0.8133, which is 0.9019.

EXAMPLE

Q. Shelbi sells two types of bread: sourdough and rye. Today she made 10 loaves of sourdough and 14 loaves of rye. In a random sample of five loaves purchased by a customer, what is the mean, variance, and standard deviation of the number of loaves that are sourdough?

A. The marked population is the sourdough loaves, so $M = 10$, $N = 24$, and $N - M = 24 - 10 = 14$. $n = 5$ is the sample size. The mean is $n\frac{M}{N} = 5\frac{10}{24} = 2.083$. The variance is $\frac{nM}{N^2(N-1)}(N-M)(N-n) =$ $\frac{5(10)}{24^2(24-1)}(24-10)(24-5) = 1.004$. The standard deviation is the square root of 1.004, which is 1.002.

---

12 At the end of the fiscal year party, Bob gives out four prizes at random to his staff of ten people, six of whom are women and four of whom are men. A person may not win more than one prize. X counts the number of women who get a prize. What is the mean, variance, and standard deviation of X?

13 Suppose that you are throwing a party for 25 people and you are giving away five prizes. Ten people at the party live in your neighborhood, and 15 people at the party live outside your neighborhood. X = the number of prizes that go to people who live in your neighborhood. What is the expected value, variance, and standard deviation of X?

---

14 Suppose that you need to randomly build a committee of 5 people from a group of 12 men and 8 women. What is the expected value, variance, and standard deviation of the number of men on the committee?

15 Suppose that you have 15 tagged fish in a tank of 25 fish. You randomly sample eight fish. What is the mean, variance, and standard deviation of the number of tagged fish in your sample?

Solutions to Problems in Not Getting Hyper about the Hypergeometric Distribution

(1) True. In the hypergeometric distribution, $M < N$. The number of marked individuals has to be less than the population size.

(2) True. The total number of individuals in the unmarked population must be $N - M$.

(3) False. Sampling one individual at a time with replacement is not the same as reaching into the population and choosing a sample all at once without replacement.

(4) This is not a hypergeometric distribution because Bob is sampling with replacement since the same person can win another prize, not without replacement, as would need to happen for it to be a hypergeometric distribution.

(5) This is a hypergeometric distribution as it meets all the criteria.

(6) X is not a hypergeometric distribution because you are sampling individuals, not a group all at once, and you are counting the number of trials, not the number of women.

(7) In this problem, $N = 11$, $M = 5$, $n = 3$, and $x = 2$, so the chance that two of the prizes go to people in your neighborhood is 0.3636:

$$P(X=2) = \frac{\binom{M}{x}\binom{N-M}{n-x}}{\binom{N}{n}} = \frac{\binom{5}{2}\binom{11-5}{3-2}}{\binom{11}{3}} = \frac{10\binom{6}{1}}{\binom{11}{3}} = \frac{10(6)}{165} = 0.3636.$$

(8) The boundaries for the domain of X are $\max(0, M+n-N) \le x \le \min(M, n) \to \max(0, 5+3-11) \le x \le \min(5,3) \to 0 \le x \le 3$.

$$P(X=0) = \frac{\binom{M}{x}\binom{N-M}{n-x}}{\binom{N}{n}} = \frac{\binom{5}{0}\binom{11-5}{3-0}}{\binom{11}{3}} = \frac{1\binom{6}{3}}{\binom{11}{3}} = \frac{1(20)}{165} = 0.1212$$

$$P(X=1) = \frac{\binom{M}{x}\binom{N-M}{n-x}}{\binom{N}{n}} = \frac{\binom{5}{1}\binom{11-5}{3-1}}{\binom{11}{3}} = \frac{5\binom{6}{2}}{\binom{11}{3}} = \frac{5(15)}{165} = 0.4545$$

$$P(X=2) = \frac{\binom{M}{x}\binom{N-M}{n-x}}{\binom{N}{n}} = \frac{\binom{5}{2}\binom{11-5}{3-2}}{\binom{11}{3}} = \frac{10\binom{6}{1}}{\binom{11}{3}} = \frac{10(6)}{165} = 0.3636$$

$$P(X=3) = \frac{\binom{M}{x}\binom{N-M}{n-x}}{\binom{N}{n}} = \frac{\binom{5}{3}\binom{11-5}{3-3}}{\binom{11}{3}} = \frac{10\binom{6}{0}}{\binom{11}{3}} = \frac{10(1)}{165} = 0.0606$$

9 The boundary conditions for X, the number of Milky Ways selected, are

$$\max(0, M + n - N) \le x \le \min(M, n) \to \max(0, 8 + 4 - 18) \le x \le \min(8,4) \to 0 \le x \le 4.$$

10 In this problem, $N = 18$, $M = 8$, $n = 4$, and $x = 4$, so $P(X = 4) = \dfrac{\dbinom{M}{x}\dbinom{N-M}{n-x}}{\dbinom{N}{n}} = \dfrac{\dbinom{8}{4}\dbinom{18-8}{4-4}}{\dbinom{18}{4}} =$

$$\dfrac{70\dbinom{10}{0}}{\dbinom{18}{4}} = \dfrac{70(1)}{3060} = 0.0229.$$

11 The problem doesn't say which type of bread is the marked population, but you want the same number from each group, so it doesn't matter. Let sourdough be the marked population. Here,

$$N = 9, M = 5, n = 6, \text{ and } x = 3, \text{ so } P(X = 3) = \dfrac{\dbinom{M}{x}\dbinom{N-M}{n-x}}{\dbinom{N}{n}} = \dfrac{\dbinom{5}{3}\dbinom{9-5}{6-3}}{\dbinom{9}{6}} = \dfrac{10\dbinom{4}{3}}{84} = \dfrac{10(4)}{84} = 0.4762.$$

12 The marked population is the women, so $M = 6$, $N = 10$, and $N - M = 10 - 6 = 4$. $n = 4$ is the sample size (prizes). The mean is $n\dfrac{M}{N} = 4\dfrac{6}{10} = 2.4$. The variance is $\dfrac{nM}{N^2(N-1)}(N-M)(N-n) = \dfrac{4*6}{10^2(10-1)}(10-6)(10-4) = 0.64$. The standard deviation is the square root of $0.64 = 0.80$.

13 The marked population is the people in your neighborhood, so $M = 10$, $N = 25$, and $N - M = 25 - 10 = 15$. $n = 5$ is the sample size (prizes). The mean is $n\dfrac{M}{N} = 5\dfrac{10}{25} = 2$. The variance is $\dfrac{nM}{N^2(N-1)}(N-M)(N-n) = \dfrac{5*10}{25^2(25-1)}(25-10)(25-5) = 1$. The standard deviation is the square root of $1 = 1$.

14 The marked population is the men, so $M = 12$, $N = 20$, and $N - M = 20 - 12 = 8$. $n = 5$ is the sample size. The mean is $n\dfrac{M}{N} = 5\dfrac{12}{20} = 3$. The variance is $\dfrac{nM}{N^2(N-1)}(N-M)(N-n) = \dfrac{5*12}{20^2(20-1)}(20-12)(20-5) = 0.9474$. The standard deviation is the square root of $0.9474 = .9733$.

15 The marked population is the tagged fish, so $M = 15$, $N = 25$, and $N - M = 25 - 15 = 10$. $n = 8$ is the sample size. The mean is $n\dfrac{M}{N} = 8\dfrac{15}{25} = 4.8$. The variance is $\dfrac{nM}{N^2(N-1)}(N-M)(N-n) = \dfrac{8*15}{25^2(25-1)}(25-15)(25-8) = 1.36$. The standard deviation is the square root of 1.36, which is 1.1662.

5

For the Hotshots: Continuous Probability Models

Identify the characteristics of the continuous uniform distribution and find probabilities for it.

Calculate the mean, variance, and standard deviation of the continuous uniform distribution.

Measure the probability density function for the exponential distribution.

Uncover the exponential distribution's relationship with the Poisson distribution.

Chapter **17**

Staying in Line with the Continuous Uniform Distribution

A *continuous random variable*, X, is a random variable that has an uncountably infinite number of possible values that fall onto the real number line. For example, X could represent the length of time spent waiting for a cab (where time can be measured to as many decimal places as you want), or the time between arrivals of planes at an airport.

The probability density function (pdf), known as f(x), of a continuous random variable doesn't give the probability of x; it tells you how dense the probability is at that point x, because the probability at any single exact point in a continuous situation is zero. So how do you find probabilities of continuous random variables? You find the area under the curve of the density function, f(x). The probability density function can have many different forms as long as the values are always nonnegative and the total area under the curve is 1.

In this chapter, you explore the characteristics of the *continuous uniform distribution*, the most basic continuous random variable. Other continuous random variables are the normal (Chapter 9) and the exponential (Chapter 18). You also practice calculating the mean, variance, and standard deviation; the density function; probabilities; and percentiles of this distribution.

Characterizing the Continuous Uniform Distribution

A continuous random variable, X, has a continuous uniform distribution when f(x) is a straight (uniform) line. The values of X go from a to b, where a and b can be any real numbers. The continuous uniform distribution is often used to measure time and comes in handy in situations when you may not have a lot of information about what the density function actually looks like. The basic shape of the continuous uniform distribution is a rectangle, since f(x) is a flat line and X goes from a to b. Let's take a look at some examples.

Suppose that X is the time you're going to wait for a cab, and it has a continuous uniform distribution where X is between 0 and 10 minutes. In this scenario, $f(x) = \frac{1}{10}$. The function looks like Figure 17-1.

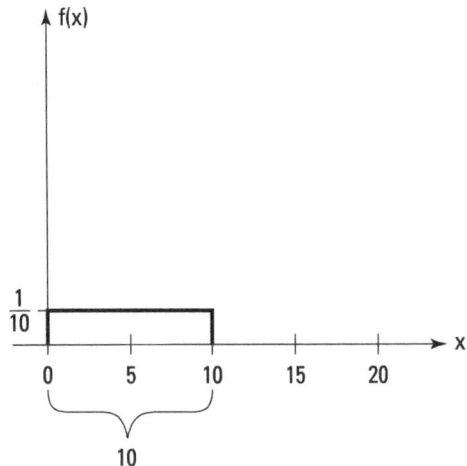

FIGURE 17-1: Looking at f(x) where X is uniform on the domain [0, 10].

REMEMBER

The *domain* of X is the set of all possible values. In a continuous uniform distribution, the domain is always an interval from a to b. Keep a and b in your mind when you are doing > or < probabilities.

You may be wondering why f(x) is $\frac{1}{10}$. It's because of the geometry of a rectangle. (That's all the geometry we're going to use here, by the way.) The area of a rectangle is calculated by multiplying length times width. The length of this rectangle is 10 − 0 = 10. That means in order for the total area to be 1 — one of the conditions of f(x) — the width, or height if you want to look at it that way, is. $\frac{1}{(10-0)} = \frac{1}{10}$. You can check the math to see that $(10-0) * \frac{1}{(10-0)} = 10 * \frac{1}{10} = 1$. So, it has to be $\frac{1}{10}$ for f(x), where X is between 0 and 10. You have f(x) = 0 everywhere else.

Suppose that you know the cab will be late, and the wait time will be a uniform distribution, where X = 10 to 20 minutes. Now, what does f(x) look like? It turns out it would look the same, just moved down the number line from 0–10 to 10–20. And f(x) would also be the same, $\frac{1}{10}$. Why? Because the formula for the length of a rectangle is (b − a), where a is the left endpoint and b is the right endpoint. The formula for f(x) is $\frac{1}{(b-a)}$, and you know $(b-a) * \frac{1}{(b-a)} = 1$.

In this example, $a = 10$ and $b = 20$, so the length is $(20 - 10)$ and the width (height) is $\frac{1}{(20-10)} = \frac{1}{10}$, where X is between 10 and 20. The picture of this situation is shown in Figure 17-2.

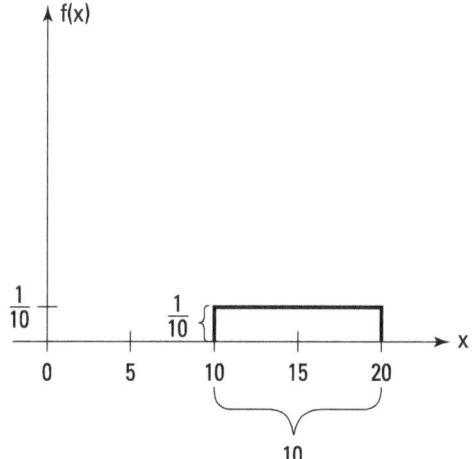

FIGURE 17-2:
Looking at f(x)
where X is
uniform on the
domain [10, 20].

The density function, f(x), has to be both non-negative and have a total area under the curve equal to 1. Keeping this in mind will be especially helpful in this chapter as you think about the continuous uniform distribution and how it looks like a rectangle.

REMEMBER

Q. You are waiting for the school bus. The wait time has a continuous uniform distribution on the interval 0 to 15 minutes (which means the possible values go from 0 to 15 minutes).

EXAMPLE

    a. What is f(x)?

    b. What is the picture of f(x)?

A. In this situation, you have $a = 0$ and $b = 15$.

    a. $f(x) = \frac{1}{(b-a)} = \frac{1}{(15-0)} = \frac{1}{15}$

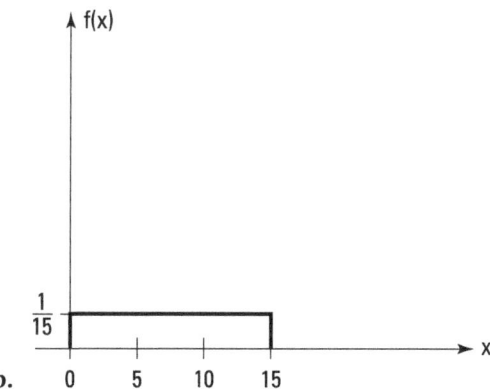

    b.

1 Tell whether the following statement is true or false: "It's okay to have a and b be negative values as endpoints on the continuous uniform."

2 Suppose that $f(x) = \frac{1}{5}$ for X, and X is a continuous uniform distribution starting at 0 and ending at b. What is the value of b?

3 Suppose that $f(x) = \frac{1}{2}$, and X starts at 0 and ends at b.

a. What is the value of b?

b. What is the picture of f(x)?

4 If X is a continuous uniform distribution on the domain [0, 10], then f(2) is what value?

5 If X is a continuous uniform distribution on the domain [0, 10], is f(2) the probability that X = 2? If not, what is the probability that X = 2?

Finding Probabilities for the Continuous Uniform Distribution

Finding probabilities for the continuous uniform distribution is unique. Usually, with a continuous distribution, you would use calculus (integration) to find the area under the curve f(x) that represents the probability. Or, as in the case of the normal distribution, you would look at a table of probabilities where the work has already been done for you. But the continuous uniform is different. The area under a flat line is a rectangle; it's just a matter of finding the area — again by taking the length times width (height).

For example, suppose that X is the time spent waiting for a cab (in minutes), which is a uniform distribution on the interval (i.e. domain) [0, 10], and you want P(X > 6). To solve this problem, first, draw a picture of the situation (see Figure 17-3). Figure 17-3 shows a shaded rectangle, and you know that the rectangle's area is the probability that X is between 6 and 10. In this case, because there is no area after 10, P(X > 6) is equal to P(6 < X < 10). That's why you draw a picture!

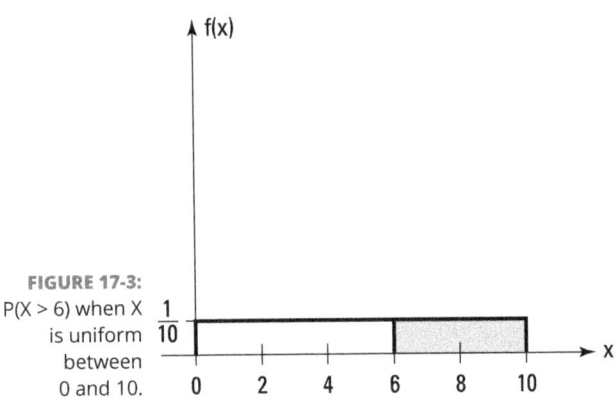

FIGURE 17-3:
P(X > 6) when X
is uniform $\frac{1}{10}$
between
0 and 10.

The probability between 6 and 10 is the length, (10 − 6) times the width (height), $\frac{1}{10}$, which gives you 4 $*$ $\frac{1}{10}$ or 0.40. The probability of waiting more than 6 minutes for a cab is 40 percent.

P(X > 6) = P(X ≥ 6) in any continuous case because the only difference between them is P(X = 6), which is 0. There is no length when you are looking at the area under the curve at one exact point. Similarly, P(X < 6) = P(X ≤ 6).

TIP

Following are the general steps for finding probabilities for a continuous uniform distribution:

1. Subtract the left endpoint from the right endpoint. This is the length.

2. Find the width (height), f(x).

3. Multiply the length and the width together.

4. Check to make sure your answer is between 0 and 1, inclusive.

For example, suppose that you want to find P(3 < X ≤ 5) when X is uniform on (2, 10). The right endpoint is 5, the left endpoint is 3, and the difference is 5 − 3 = 2. This is the length. The next step is to find the width, f(x). This is $\frac{1}{(b-a)} = \frac{1}{(10-2)} = \frac{1}{8}$. Now multiply the length and the width to get 2 $*$ ⅛ = 0.25. You can see what's going on in Figure 17-4.

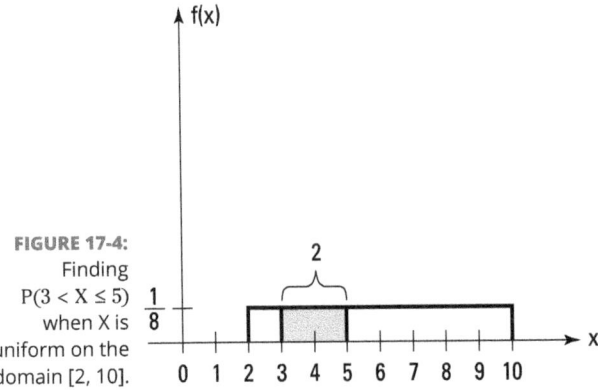

FIGURE 17-4:
Finding
P(3 < X ≤ 5) $\frac{1}{8}$
when X is
uniform on the
domain [2, 10].

EXAMPLE

Q. Suppose that X is uniform on (0, b) and f(x) = ½. Find P(¼ < X < ¾).

A. First, find b. You have (b − 0) * ½ = 1, so $\frac{b}{2} = 1$ and b = 2. So, X is uniform on (0, 2). Then P(¼ < X < ¾) is $(¾ − ¼) * ½ = \frac{2}{4} * ½ = ¼ = 0.25$.

---

6 Suppose that X is uniform on [0, 1/2]. Find P(0 < X < ¼).

7 Tell whether the following statement is true or false: "If X is uniform on [0, 10], then $P(0 < X < 5)$ is less than $P(0 \le X \le 5)$."

---

8 Suppose that X is uniform from 5 to b. If $P(5 < X < 8) = ½$, what is b?

9 Suppose that X is uniform from [−5, 5]. What is the probability that X is between −2 and 2?

Calculating the Mean, Variance, and Standard Deviation of the Continuous Uniform Distribution

The continuous uniform distribution is symmetric; when you cut it down the middle, it looks exactly the same on each side. That means the mean is the midpoint of the domain of X. For example, if X is uniform on (0, 10), you know that the mean is $\frac{(0+10)}{2} = 5$. In general, if X is uniform on (a, b), then the mean, μ, is $\frac{(a+b)}{2}$.

For example, if X is uniform on (6, 20), the mean is $\frac{(6+20)}{2} = \frac{26}{2} = 13$. You can see what is going on in Figure 17-5.

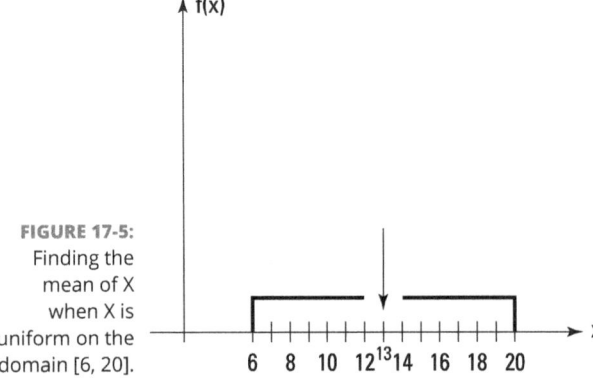

FIGURE 17-5: Finding the mean of X when X is uniform on the domain [6, 20].

TIP

The mean has to be a value in the domain of X, so make sure your answer falls somewhere between a and b.

The variance of the continuous uniform is given by the following formula:

$$\sigma^2 = \frac{(b-a)^2}{12}$$

where b is the right endpoint of the domain and a is the left endpoint of the domain. The variance is the square of the standard deviation, so the standard deviation of the continuous uniform is $\sigma = \sqrt{\frac{(b-a)^2}{12}}$. For example, if X is uniform on [0, 10], the variance of X is $\sigma^2 = \frac{(10-0)^2}{12} = \frac{100}{12} = 8.33$, and the standard deviation is $\sigma = \sqrt{\frac{(10-0)^2}{12}} = \sqrt{8.33} = 2.89$.

REMEMBER

The variance and standard deviation are both measures of variability, and both must be non-negative. And it makes sense that when a and b are further apart, there will be more variability in X, as shown in the formula (b − a)² will be a bigger value as well.

Q. Suppose that X is uniform on (2, 10). Find the mean, variance, and standard deviation of X.

A. The formula for the mean of X is $\frac{(a+b)}{2} = \frac{(2+10)}{2} = \frac{12}{2} = 6$. The variance is $\sigma^2 = \frac{(10-2)^2}{12} = \frac{64}{12} = 5.33$. The standard deviation is the square root of 5.33, which is 2.31.

---

10 Suppose that X is uniform on (0, ½). Find the mean, variance, and standard deviation of X.

11 Suppose that X is uniform on (−10 to 10). Find the mean, variance, and standard deviation of X.

12 Suppose that X is uniform from 5 to b.

    a. If the mean is 10, what is b?

    b. What are the variance and standard deviation in that case?

13 Is the standard deviation of X always less than or equal to the variance of X?

Calculating Percentiles for the Continuous Uniform Distribution

A percentile is a value of X that has a certain percentage of the values below it. For example, if X is at the 50th percentile, then 50 percent of the values lie below X. (X would be at the median, and in the case of the continuous uniform, X would be the midpoint as well.)

So, in a way, you work backward to find percentiles for X in a continuous uniform distribution. Pictures help as well. Let's assume that you have 10 minutes in which to complete a task at work, and X is the time it takes you to do it. You finish the task in 8 minutes. At what percentile did you finish? In other words, what percentage of other times lie below yours? Take a look at Figure 17-6. To find the percentile when X is 8 you find the area below 8. That means finding the distance below 8, which is $8 - 0 = 8$, divided by the total distance (which is $10 - 0 = 10$). You can see that $\frac{8-0}{10-0} = 0.80$, so 80 percent of the data lies below 8, and 8 minutes is at the 80th percentile.

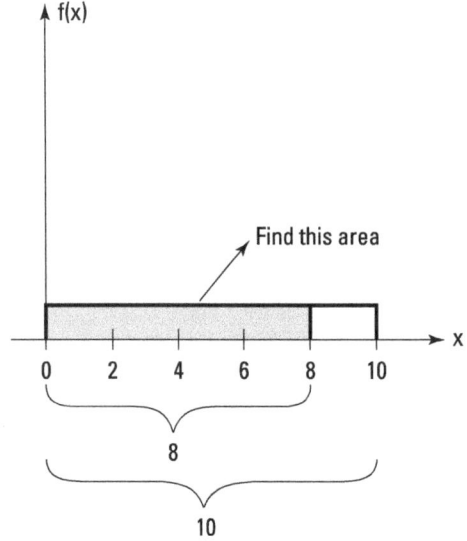

Now let's change the endpoints. Suppose that X is uniform on (2, 10). What percentile are you at if you are at 8? Figure 17-7 illustrates the scenario. You can see that $\frac{(8-2)}{(10-2)} = \frac{6}{8} = 75$ percent of the data lies below 8, so 8 is at the 75th percentile.

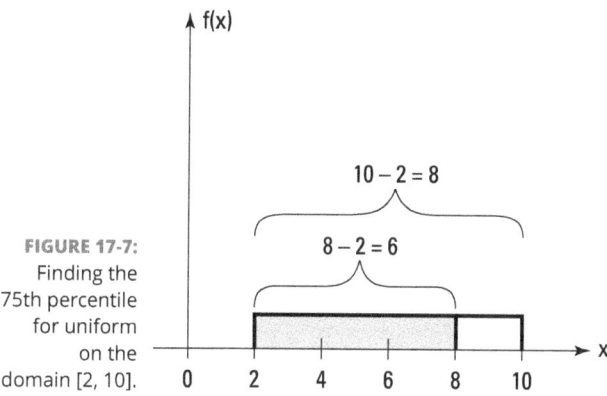

FIGURE 17-7:
Finding the
75th percentile
for uniform
on the
domain [2, 10].

In general, to what percentile you are at for a number x between a and b, do the following:

1. Draw a picture of the situation.

2. Find (x – a) where a is the left endpoint.

3. Find (b – a) where a is the left endpoint and b is the right endpoint.

4. Take $\frac{(x-a)}{(b-a)}$ and convert it to a percentage. This tells you what percentile you are at for x.

EXAMPLE

Q. Suppose that X is uniform on (–3, 5). Find the 25th percentile for X.

A. In this case, you know the percentage, so you find where it lies. The following picture shows us the situation. You know the 25th starts at –3, but you don't know where it ends, so call it x. The 25th percentile has a length of (x – –3) and is divided by the entire length, which is 5 – –3 = 8. So, you have $\frac{(x+3)}{8} = 0.25$, which means x + 3 = 2 so x = –1. That's your 25th percentile.

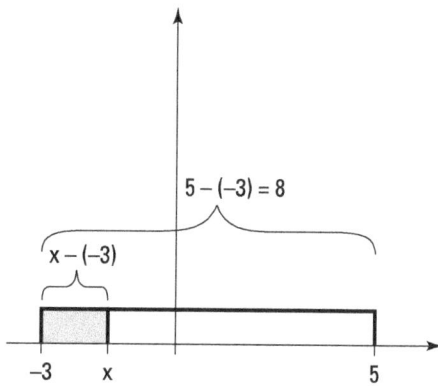

14 Suppose that X is uniform on (0, ½). Find the 50th percentile of X.

15 Suppose that X is uniform on (-10 to 10). Find the 25th percentile of X.

16 Suppose that X is uniform from 5 to b. If the 25th percentile is 8, what is b?

17 Tell whether the following statement is true or false: "The percentile is always a value between a and b in the continuous uniform distribution."

18 Explain why X being on the domain [0, 10] is the same as X being on the domain (0, 10) if X is uniform.

Solutions to Problems in Staying in Line with the Continuous Uniform Distribution

1. True. The values of X can be negative, but the values of f(x) must be nonnegative.

2. The answer is 5. You know $f(x) = 1/5$ for X, and X is a continuous uniform distribution starting at 0. You know that $(b - a) * 1/5 = 1$, and you know $a = 0$. So, you have $(b - 0) * 1/5 = 1$ and $b/5 = 1$, so $b = 5$.

3. The answer is $b = 2$. Suppose that $f(x) = 1/2$, and X starts at 0 and ends at b.

   a. You need to find b. You know $(b - 0) * 1/2 = 1$ so $b * 1/2 = 1$ and $b = 2$.

   b. The picture is shown here:

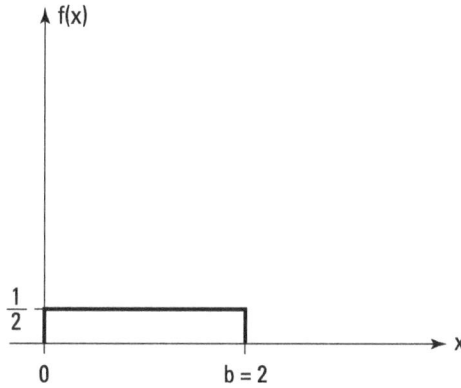

4. The answer is $1/10$. The values of f(x) are $\frac{1}{(10-0)} = \frac{1}{10}$ for all X between 0 and 10, including 2.

5. No. f(2) shows the density of the probability at 2, not the probability itself. There is no probability at any single point in the domain of X. So, the probability is 0.

6. The answer is 1/2. You take $\dfrac{\left(\dfrac{1}{4} - 0\right)}{\left(\dfrac{1}{2} - 0\right)} = \dfrac{\dfrac{1}{4}}{\dfrac{1}{2}} = \dfrac{1}{4} * \dfrac{2}{1} = \dfrac{1}{2}$.

7. False. It doesn't matter whether the endpoints are included. $P(X = 0) = 0$ and $P(X = 5) = 0$, so $P(0 < X < 5)$ is equal to $P(0 \leq X \leq 5)$.

8. The answer is 11. Suppose that X is uniform from 5 to b. If $P(5 < X < 8) = 1/2$, you have $\dfrac{(8-5)}{(b-5)} = \dfrac{1}{2}$, so $\dfrac{3}{(b-5)} = \dfrac{1}{2}$. Then $b - 5 = 6$ so $b = 11$.

9. The answer is 0.40. Suppose that X is uniform from (-5 to 5). $P(-2 < X < 2) = \dfrac{(2--2)}{(5--5)} = \dfrac{4}{10} = 0.40$.

10. The answers are $1/4 = 0.25$, $1/48 = 0.021$, and $\sqrt{1/48} = 0.14$. X is uniform on (0, 1/2). The mean is $\dfrac{(a+b)}{2} = \dfrac{(0+1/2)}{2} = \dfrac{1}{4}$. The variance is $\dfrac{(b-a)^2}{12} = \dfrac{(1/2-0)^2}{12} = \dfrac{1/4}{12} = \dfrac{1}{48}$, and the standard deviation is the square root of $1/48$, which is 0.14.

(11) The answers are 0, 33.33, and $\sqrt{33.33} = 5.77$. X is uniform on (-10 to 10); the mean is $\frac{(10 + -10)}{2} = 0$; the variance is $\frac{(10 - -10)^2}{12} = 33.33$, and the standard deviation is the square root of 33.33, which is 5.77.

(12) Suppose that X is uniform from 5 to b.

a. The answer is b = 15. The mean is $\frac{(a + b)}{2} = \frac{(5 + b)}{2} = 10$, which means $5 + b = 20$ and $b = 15$.

b. The answers are 8.33 and 2.89. The variance is $\frac{(15 - 5)^2}{12} = 8.33$. The standard deviation is the square root of 8.33, which is 2.89.

(13) No. The standard deviation is the square root of the variance, and is less than the variance when the variance is > 1. It's equal to the variance when the variance is 1, and it's greater than the variance when the variance is < 1. For example, if the variance is ¼, the standard deviation is ½, which is greater than ¼.

(14) The answer is ¼. Suppose that X is uniform on (0, 1/2). The 50th percentile of X is such that $\frac{(x - 0)}{(\frac{1}{2} - 0)} = 0.50$ so $\frac{x}{\frac{1}{2}} = 0.50$ and x = 0.25 or ¼.

(15) The answer is -5. Suppose that X is uniform on (-10 to 10). The 25th percentile of X is such that $\frac{(x - -10)}{(10 - -10)} = 0.25$, so $\frac{(x + 10)}{20} = 0.25$, x + 10 = 5, and x = -5.

(16) The answer is 17. Suppose that X is uniform from 5 to b. The 25th percentile is 8, so $\frac{(8 - 5)}{(b - 5)} = 0.25$. That means $\frac{3}{(b - 5)} = \frac{1}{4}$, so b - 5 = 12 and b = 17.

(17) True. A percentile is always a value between a and b. It must be a value in the domain of X.

(18) These two domains are basically the same. There is no probability at either endpoint, so it doesn't matter if you include them or not.

Chapter **18**

Exposing the Exponential Distribution (and Its Relationship to Poisson)

The exponential distribution has a probability density function (pdf) that is in the shape of the exponential function. It crosses the Y axis at a positive value called λ and is skewed to the right, heading to 0 as the X values go to infinity. It's based on the same function that models death rates and exponential decay, e to a negative power of x. Other real-world examples that call for the exponential model include lifetimes of products, times between occurrences, and the time you spend waiting in line.

In this chapter, you work on problems to find the actual pdf for the exponential distribution, see how different values of λ affect it, and find probabilities for it. You also explore the mean, variance, and standard deviation, as well as the exponential's relationship to the Poisson distribution (which is discussed in Chapter 13).

Characterizing the Exponential Distribution

Exponential distributions are used when you are modeling the passage of time, such as the time between customers or the time until an event occurs. The density function for the exponential is $f(x) = \lambda e^{-\lambda x}$, where $x \geq 0$. λ is a constant that can change with each problem; this is called a *parameter* for the density function. Notice that when $x = 0$, $f(x) = \lambda e^{-\lambda(0)} = \lambda(1) = \lambda$. So, the density function crosses the Y axis at λ, and goes downhill from there. If it crosses at a large value of λ, it drops more quickly than if it crosses at a smaller value of λ (see Figure 18-1). The total area under the curve must be one.

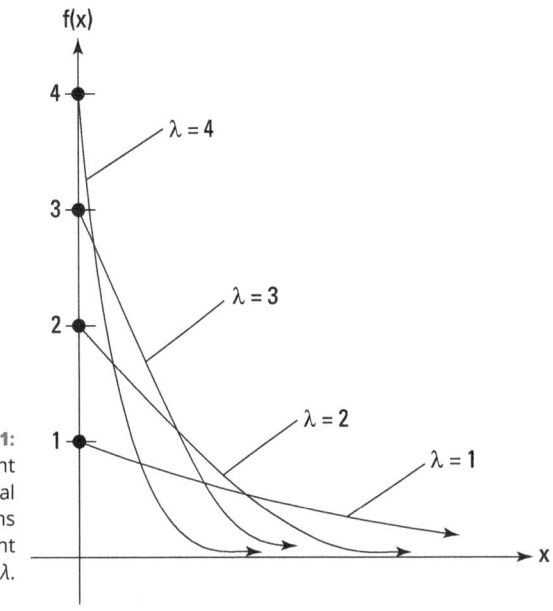

FIGURE 18-1: Different exponential distributions for different values of λ.

TIP

λ can be any number > 0. Not just numbers like 1, 2, 3, 4.

EXAMPLE

Q. What is the density function for an exponential distribution with $\lambda = \frac{1}{2}$?

A. $f(x) = \frac{1}{2}e^{-\frac{1}{2}x}$, $x \geq 0$.

1. Which exponential distribution drops faster in the graph, the one with $\lambda = 10$ or the one with $\lambda = \frac{1}{10}$?

2. Tell whether the following statement is true or false: "The exponential function eventually goes below the x-axis."

3. Fill in the blank with the best answer: A _____ is a constant in a density function that can change with each problem.

4. Where does the exponential density function cross the y-axis?

Finding Probabilities for the Exponential Distribution

In this section, you go through each of the different types of probabilities and figure out how the exponential distribution finds those probabilities. You look at less-than, greater-than, and between two values.

REMEMBER Because the exponential distribution is continuous, there is no probability at an exact point. For example, $P(X = 2) = 0$. This also means $P(X < 2) = P(X \leq 2)$ because there is no probability right at 2, so you get the same answer whether you include it or not. Same for greater-than: $P(X > 2) = P(X \geq 2)$. For the between probabilities, you can include or not include either endpoint and you'll get the same answer. So, we will just work with <, >, and a $< X <$ b.

Less-than (or less-than-or-equal-to) probabilities

$P(X < x)$ is not found on a table with the exponential distribution. Mathematicians have done the work of finding a function that you put x into (involving calculus), and it gives you this probability. It's called the *cumulative distribution function* (cdf), and it is noted by $F(x)$. The function $F(x)$ for the exponential distribution is $1 - e^{-\lambda x}$ for any value of $x > 0$ that you want to find $P(X < x)$ for.

For example, if you want $P(X < 1)$, and $\lambda = 4$, you calculate $1 - e^{-(4)\,(1)} = 1 - 0.018 = 0.982$. If you want $P(X < \frac{1}{2})$ and $\lambda = 4$, you get $1 - e^{-(4)\,(1/2)} = 0.865$. See Figure 18-2 for $P(X < \frac{1}{2})$ when $\lambda = 4$.

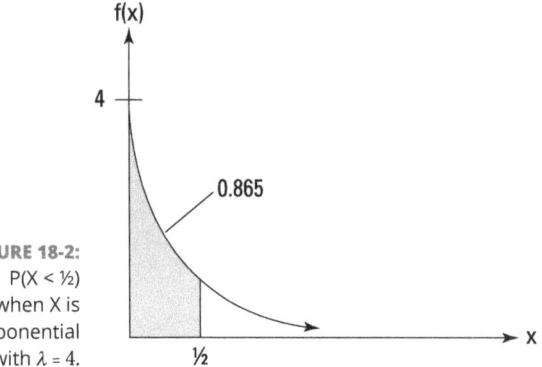

FIGURE 18-2: $P(X < \frac{1}{2})$ when X is exponential with $\lambda = 4$.

Greater-than (or greater-than-or-equal-to) probabilities

Because greater-than is the complement of less-than (without worrying about the equal to, which has a probability of zero), the function you use to find $P(X > x)$ is $1 - P(X < x) = 1 - (1 - e^{-\lambda x}) = e^{-\lambda x}$. So if you want $P(X > 1)$ and $\lambda = 4$, you have $e^{-\lambda x} = e^{-4(1)} = 0.018$; and if you want $P(X > \frac{1}{2})$ and $\lambda = 4$, you get $e^{-\lambda x} = e^{-4(1/2)} = e^{-2} = 0.135$. (Notice these are the complements of the probabilities found under the less-than section.) See Figure 18-3 for $P(X > \frac{1}{2})$ when $\lambda = 4$.

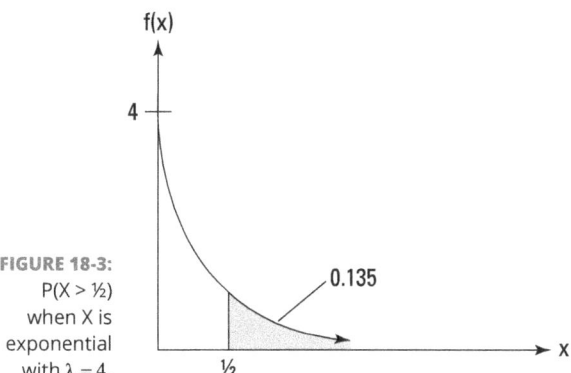

FIGURE 18-3:
P(X > ½)
when X is
exponential
with λ = 4.

Probability between two values (including the endpoints or not)

To find the probability between two values, a and b, with a continuous distribution like the exponential, you take the probability all the way up to b and subtract the probability all the way up to a. In other words, you take $P(X < b) - P(X < a)$. Looking at the less-than section from previous work, $P(X < b) = 1 - e^{-\lambda b}$ and $P(X < a) = 1 - e^{-\lambda a}$, so when you subtract them, you get $(1 - e^{-\lambda b}) - (1 - e^{-\lambda a}) = 1 - e^{-\lambda b} - 1 + e^{-\lambda a} = e^{-\lambda a} - e^{-\lambda b}$. So, if you want to find $P(\frac{1}{2} < X < 1)$ when $\lambda = 4$, you take $e^{-\lambda(1/2)} - e^{-\lambda(1)} = e^{-4(1/2)} - e^{-4(1)} = e^{-2} - e^{-4} = 0.117$. Figure 18-4 illustrates the situation.

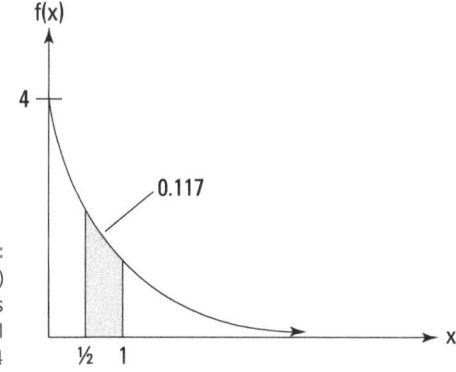

FIGURE 18-4:
P(½ < X < 1)
when X is
exponential
with λ = 4

TIP

As for any probability, always check to make sure your answer is between 0 and 1. With calculations like the ones in this section, it's easy to make an error; checking your answer helps to know if you are in the right ballpark at least.

EXAMPLE

Q. Suppose that X is the time (in minutes) between phone calls at a help desk, which is exponential with λ = 5. What is the probability that a call is exactly 2½ minutes long?

A. Zero. There is no probability at a single exact value with a continuous distribution like the exponential. You need X to represent an interval on the number line to get a probability, like $X > 2$, $X < 2$, or $2 < X < 3$.

5. Suppose that X is the amount of time a battery lasts, and is exponential with $\lambda = \frac{1}{50}$.

   a. What's the chance that the battery lasts less than 50 hours?

   b. What's the chance that the battery lasts more than 60 hours?

   c. What's the chance that the battery lasts between 50 and 60 hours?

6. Bernardo wants to know the chance he's going to have to wait longer than 15 minutes for a cab to arrive. The time to wait for a cab is X, which is exponential with $\lambda = \frac{1}{20}$.

7. Suzette wants to know the chance she's going to have to wait less than 15 minutes on the phone to get a customer associate on the helpline, where X = the time to wait for an associate (in minutes), which is exponential with $\lambda = \frac{1}{30}$.

8. Erica wants to know the chance that her candle will burn between 10 and 15 hours, where X = the candle's lifetime, which is exponential with $\lambda = \frac{1}{20}$.

Expressing the Mean, Variance, and Standard Deviation of the Exponential

The mean of the exponential is $\frac{1}{\lambda}$. So if X is the amount of time a phone call lasts on a helpline, and X is exponential with a mean of 10 minutes, then the density function is $f(x) = \lambda e^{-\lambda x} = \left(\frac{1}{10}\right)e^{-\left(\frac{1}{10}\right)x}$. And if another exponential distribution has the density function $4e^{-4x}$, then the mean is ¼.

REMEMBER

This is a very important point. The mean is not λ; it's $\frac{1}{\lambda}$. (With a Poisson distribution, the mean is λ, but not so for the exponential.)

The variance of the exponential is $\frac{1}{\lambda^2}$, and the standard deviation is the square root of the variance, which is $\frac{1}{\lambda}$. The standard deviation is in the units of X. (The variance is in square units, which is usually not interpretable.)

So if the time between arrivals of phone calls (in hours) is exponential with $\lambda = 10$, then the mean is $\frac{1}{10}$ of an hour, the variance is $\frac{1}{10^2} = \frac{1}{100}$, and the standard deviation is $\frac{1}{10}$ of an hour.

TIP

You can change $\frac{1}{10}$ hour to minutes by remembering this proportion: $\frac{\text{min}}{\text{hour}} = \frac{60}{1} = \frac{x}{1/10} \rightarrow 60(1/10) = 1x \rightarrow x = 6$ minutes.

Suppose that X is the number of minutes waiting for a cab, which is exponential with the probability density function, $(\frac{1}{10})e^{-1/10(x)}$, where $x \geq 0$. We have $\lambda = \frac{1}{10}$, so the mean is $\frac{1}{\frac{1}{10}} = 10$ minutes, the standard deviation is 10 minutes, and the variance is $10^2 = 100$.

EXAMPLE

Q. Suppose that X is the time (in hours) between phone calls at a help desk, which is exponential with $\lambda = 5$. What is the mean and standard deviation of X?

A. The mean is $\frac{1}{\lambda} = \frac{1}{5}$ of an hour. So is the standard deviation. This is 12 minutes since $\frac{\text{min}}{\text{hour}} = \frac{60}{1} = \frac{x}{\frac{1}{5}} \rightarrow 60(\frac{1}{5}) = 1x \rightarrow x = 12$.

---

 9 Suppose that Stacy is waiting for a cab, and X = the minutes she spends waiting, which has an exponential distribution with a mean of 10 minutes. What is the density function?

 10 Suppose that Mildred bought some batteries that are supposed to last 100 hours on average. What's the standard deviation of the battery lifetime?

11 Frankie works at a helpline. He notices that the time between calls has a large amount of variance, so the standard deviation must be large. What must be true about the time between calls?

a. λ must be large

b. $\frac{1}{\lambda}$ must be large

 12 Suppose that X = the length of time (in minutes) between phone calls to a helpline, which has an exponential distribution with a density function of $f(x) = 3e^{-3x}$. What's the mean and standard deviation of X?

Relating the Exponential and Poisson Distributions

It's not a coincidence that we used the same name for the parameter, λ, in the Poisson and the exponential distributions. These two distributions are related. The Poisson distribution (discussed in Chapter 13) models the *number* of arrivals or occurrences in a fixed period of time. The exponential distribution models the *time between* those arrivals or occurrences.

Suppose that X represents the number of people who arrive at a shoe store in an hour and has a Poisson distribution with λ (the mean) equal to 10. So, the mean of this Poisson distribution is ten arrivals per hour. What do you think the mean time between customers is? The time between customers is exponential. It makes sense, since ten arrivals per hour means on average one every 6 minutes ($6 * 10 = 60$ minutes $= 1$ hour).

REMEMBER

In general, if X is a Poisson distribution and is equal to the number of arrivals in a fixed period of time and has a mean of λ, the time between arrivals, Y, has an exponential distribution with a mean of $\frac{1}{\lambda}$. It's the same λ in both distributions; it's just flipped over in the exponential.

TIP

You just have to make sure you have the time periods right; you need to stay consistent throughout. One-tenth of an hour is 6 minutes. Remember the conversion equation: $\frac{\text{min}}{\text{hour}} = \frac{60}{1} = \frac{x}{\frac{1}{10}} \rightarrow 60(\frac{1}{10}) = 1x \rightarrow x = 6$ minutes.

For example, suppose that X is the number of returns that come into the customer service desk per hour. X is a Poisson distribution with a mean of 15. What is the distribution of the time between returns? Y = the time between returns, and is exponential with a parameter of $\frac{1}{15}$ hour, which equals 4 minutes since $\frac{min}{hour} = \frac{60}{1} = \frac{x}{\frac{1}{15}} \rightarrow 60(\frac{1}{15}) = 1x \rightarrow x = 4$ minutes.

Now, to pull it all together, Table 18-1 compares the Poisson and exponential distributions.

Table 18-1   The Poisson and Exponential Distributions

	Probability Mass/ Density Function	Mean	Var.	Std Dev
Poisson	$\frac{e^{-\lambda x}}{x!}, x = 0,1,2\ldots$	λ	λ	$\sqrt{\lambda}$
Exponential	$\lambda e^{-\lambda x}, x \geq 0$	$\frac{1}{\lambda}$	$\frac{1}{\lambda^2}$	$\frac{1}{\lambda}$

Q. Suppose that X is the time (in hours) between phone calls at a help desk, which is exponential with $\lambda = 5$. How many phone calls do you expect in an hour?

EXAMPLE

A. X is exponential and the mean of X is $\frac{1}{\lambda}$, which equals $\frac{1}{5}$ of an hour. The number of arrivals per hour is Poisson with mean $\lambda = 5$.

---

13 Suppose that you know the number of cabs that come per hour has a Poisson distribution with a mean equal to 6. Therefore, the minutes spent waiting for a cab, Y, has what distribution and what mean?

14 Suppose that Gordie notices the number of sales on his Internet website is two sales per hour on average, and it has a Poisson distribution. How much time on average is there between sales? Name the distribution and the average in hours and minutes.

15 Frankie works at a helpline. He notices that about five calls per hour come in on average. What is the average time between calls? Name the distribution and the average.

16 If X = the length of time (in minutes) between phone calls to a very busy helpline dispatcher, and X has an exponential distribution with density function $f(x) = 3e^{-3x}$, what's the mean and standard deviation of Y if Y = the number of phone calls that arrive in one minute? In one hour?

Solutions to Problems in Exposing the Exponential Distribution (and Its Relationship to Poisson)

(1) The answer is the one with $\lambda = 10$. A higher value of λ crosses the y-axis at a higher point, and then drops faster than a lower value of λ.

(2) False. The exponential density function always remains nonnegative; it's one of the rules of any density function.

(3) A <u>PARAMETER</u> is a constant in a density function that can change with each problem.

(4) The answer is λ. The exponential density function cross the y-axis at the point $(0, \lambda)$.

(5) (a.) The chance that the battery lasts less than 50 hours is $1 - e^{-(1/50)(50)} = 1 - e^{-1} = 0.632$.

(b.) The chance that the battery lasts more than 60 hours is $e^{-(1/50)(60)} = e^{-1.2} = 0.301$.

(c.) The chance that the battery lasts between 50 and 60 hours is $e^{-(1/50)50} - e^{-(1/50)60} = 0.067$.

(6) The answer is 0.472. The time to wait for a cab in minutes is X, which is exponential with $\lambda = \frac{1}{20}$. We want the time to wait longer than 15 minutes for a cab, so we want $P(X > 15) = e^{-(1/20)(15)} = e^{-0.75} = 0.472$.

(7) The answer is 0.607. The time to wait on a call (in minutes) is X, which is exponential with $\lambda = \frac{1}{30}$. We want the time to wait to be less than 15 minutes, so we want $P(X < 15) = 1 - e^{-1/30(15)} = 1 - e^{-0.50} = 1 - 0.607 = 0.393$.

(8) The answer is 0.134. X = the candle's lifetime in hours, and is exponential with $\lambda = \frac{1}{20}$. Therefore, we want $P(10 < X < 15) = e^{-(1/20)(10)} - e^{-(1/20)(15)} = 0.134$.

(9) The density function is $f(x) = (1/10)e^{-(1/10)(x)}$, $x > 0$.

(10) The answer is 100 hours. The standard deviation of the battery lifetime is the same as the mean, 100 hours.

(11) The answer is (b.) $1/\lambda$ must be large.

(12) Since $\lambda = 3$, the mean and standard deviation are both $1/\lambda$, which equals $1/3$ of a minute, which is 20 seconds.

(13) The exponential distribution and the mean is 1/6 of an hour, which is 10 minutes.

(14) The distribution is exponential, and the average is 1/2 hour, or 30 minutes.

(15) The distribution is exponential, and the average is 1/5 hours, which equals 12 minutes.

(16) Y is a Poisson distribution with a mean of $\lambda = 3$ calls per minute, and $60(\lambda) = 180$ calls per hour.

6

The Part of Tens

Identify common pitfalls when calculating probability and how to avoid them.

Understand the chance of winning the lottery and why there's no such thing as being "on a roll" when playing casino games.

Refresh your understanding of the ten probability distributions discussed in this book.

Chapter **19**

Top Ten Probability Mistakes

Probability may seem intuitive on the surface — and some of it is. However, some of it isn't. Many mistakes can be made if you don't understand what's really going on behind the scenes. In this chapter, you take a look at some of the most common probability errors and misconceptions that can happen to anyone, just because of the nature of probability. It's not the easiest subject in the world; however, clearing up common pitfalls will help you raise the bar on your level of understanding and help you avoid problems down the road.

Forgetting Where Probabilities Live

Probabilities have strict rules, and you always need to check when you are finished with a problem to make sure those rules are being followed. The two biggest rules of probability are:

» Each probability lies between 0 and 1.

» The sum of all the probabilities of all the outcomes in the sample space, S, must be 1.

A famous statistician I know has a framed picture in his office of a problem from one of the exams he took in a high-level probability class. He worked the problem feverishly; you can see on the page that the amount of work he put into it is stunning. When he was finished, he got an answer of 2 for the probability. He wiped his brow, took a deep breath, and merrily moved on to the next problem *without checking* the most important rule of all. His teacher wasn't impressed and gave him a zero for that problem. Let's make sure the same doesn't happen to you!

If you haven't already done so, get into the habit of checking every probability you calculate to make sure it falls between 0 and 1. Zero means it could never happen, and 1 means it's guaranteed to happen. Most probabilities fall between those values.

Also be sure to check your answers to questions where the solution is something that's not a clean fraction like ½. For example, if your answer is the square root of 2, it's not valid because when you put it in your calculator, the square root of 2 is 1.414. That's larger than 1. You also can't get negative numbers for probabilities; if you do, a calculation mistake was made along the way. Make sure you've got the right formula and have worked out all the details correctly.

Remember the order of operations, PEMDAS, to avoid issues: P = parentheses, E = exponents, M = multiplication, D = division, A = addition, and S = subtraction.

Misinterpreting Very Small Probabilities

A very small probability could be something like 0.00000002. It can be so small that you might get the idea that it could never occur. However, a big difference exists between small numbers that are close to zero and zero itself. Zero means the outcome or event could never happen, no matter what. And a probability that is very small and close to zero means the outcome is rare. But rare things do happen. People get rare forms of a disease, for example. You just can't predict who and where it will happen.

For example, if in a city an unusually large number of people develop a rare disease, you might want to connect it to a problem in the city, such as the water supply or the air pollution. Now, maybe something is wrong, but it could just be a rare event. Think about the lottery; the chances of winning are so small that it's incredible, yet someone is going to win eventually. Yes, it may be you, because it has to be somebody, but probably not. You can't predict small probability events because of the rate at which they occur. Yet they will occur eventually. Even a probability of 0.00000002 means 2 out of 100,000,000 times it's expected to happen over the long term (plus or minus). That's not zero.

Using Probability for Short-Term Predictions

Probability is a long-term phenomenon; you can't use it to predict short-term events with any consistency. For example, take the roll of a die. The probability of each outcome is ⅙, but that doesn't give you any information as to which outcome will show up next, a 1 or a 2 or a 3 or a 4 or a 5 or a 6. So you pick a number, and maybe you're right, maybe you're wrong. But the laws of probability say that no matter what you think, you're only going to be right one-sixth of the time in the long term. This makes betting difficult.

When you roll a fair die, all outcomes are equally likely. However, if the die is not fair, certain outcomes have higher or lower probabilities. A higher probability doesn't guarantee that an outcome is going to come up first; a lower probability doesn't guarantee it's never going to come up. For example, roll two dice and look at the sum. A sum of 7 has the highest probability because, among the combinations of two dice, there are many that add up to 7: 1 and 6, 6 and 1, 2 and 5, 5 and 2, 3 and 4, and 4 and 3. Out of the $6 * 6 = 36$ outcomes of rolling two dice, 6 of them sum to 7. But that doesn't mean a 7 is "due" if you haven't seen it for a while. All outcomes are independent, which means they don't affect each other. It just means that in the long term, you should see a sum of 7 most often. But that's the long term. That's why, when you do pick the outcome correctly and win some money, you should take your money and stop.

Before you start gambling, set limits for how much you are willing to lose or how much you'd be happy winning. Because remember: Probabilities are set up in casinos so that in the long term, the casino will win.

Thinking 1-2-3-4-5 and 6 Can't Win Powerball

Powerball is a multi-state lottery game in the United States where, at the time of this writing, you pick five numbers from a set of 69 white balls, and you pick one number from a set of 26 red "Powerballs." You can choose your own numbers or have the lottery terminal randomly pick them for you. Both options give you the same chance of winning because all tickets have the same chance of winning; the results are independent and don't affect one another. So picking your children's birth dates might make you feel lucky, but it's the same as having the random number generator decide on the numbers for you. The chance of winning the Powerball lottery is 1 in 292,201,338, regardless of what numbers you choose.

So, what's the point? The point is, you can outsmart the lottery a little bit at least. Choose numbers that no one else will pick because *they* think they could never hit, and if they do hit, you won't have to split the winnings with anybody; you can have it all to yourself. For example, take numbers 1-2-3-4-5 and number 6 for the Powerball. Nobody thinks that would win, and if it did, many people would think the system was rigged. Yet, that combination has the same chance as 15, 28, 2, 67, 43, and 6, or whatever numbers anyone else might try. It just doesn't "look" random. This shows us all how hard it is to win the lottery.

If you pick lottery numbers, pick a set of consecutive numbers so that there is a smaller chance of splitting the winnings with anyone who picked the same numbers!

Thinking "I'm on a Roll!"

Suppose that you are playing the roulette wheel and you're betting on whether the result will be a red number or a black number. So far, you've won three times in a row betting on black, so you tell yourself you're going to continue betting on black because it's lucky and you're "on a roll." That means you believe the odds are in your favor if you pick black over red. You're wrong.

REMEMBER

One thing about random phenomena like spinning wheels, flipping coins, or repeatedly playing a game is that results are *independent* (unless you're a card counter, but that's another story). That means the results don't influence each other. You can step back after the fact and say, "Wow, I got three black ones in a row!" and think you should bet on black. But you would have the same chance of getting red. (And don't forget that there is either one or two green numbers on the roulette wheel that, if you land on one, causes you to lose automatically.) Probability is a long-term process, and results are independent.

If you flip a coin ten times in your mind and write down the results without actually doing any flipping, you probably won't write down results like THHHHHHHHT. The group of eight heads is called a *run*. This result has the same chance of occurring as HTHTHTHTHTHT if you consider it before the fact. Now, after the fact, you can ask, "What are the chances of my getting eight heads in a row?" It would be ½ to the eighth power. But that's the same as getting any combo of eight values because each roll has a ½ chance of occurring. That's part of what makes probability counterintuitive at times.

REMEMBER

With slot machines, you don't know what's going on behind the random number generator that creates the outcomes. It could be that someone programmed it to be on a roll of winning; it could be that, like the coin flips, runs just happen sometimes. But pouring more money into a machine just because it seems to be winning a few times in a row might lead to trouble. Eventually, the law of large numbers wins out if you play long enough. And those odds are not in your favor.

Giving Every Two Outcomes a 50–50 Chance

You've heard it before; I've heard it before — people saying, "Well, it's a 50–50 chance of whether it happens or not." They do this when there are two outcomes, such as your friend making a free throw, or the stoplight being red, or any situation that's going to either happen or not happen. That's not exactly right. Yes, if you are flipping a fair coin, the results will be 50 percent heads and 50 percent tails, and you can say that there is a 50–50 chance of one or the other occurring. But not all situations that have two outcomes are 50–50. Not all outcomes are equally likely; in fact, most of them aren't.

For example, think about free throws in basketball. A professional NBA player has much more than a 50 percent chance of making a free throw. Maybe you do too! At any rate, it's not likely to be exactly 50–50. Traffic lights are usually red, less often than not, to keep traffic moving. It's easy to think of scenarios that at first blush seem to be a 50–50 situation just because there are two outcomes, but unless it's modeled like a fair coin (and even coins aren't *exactly* fair), it's not going to be 50–50. (Studies have shown that when accounting for certain real-world factors, coins tend to land on the side they started on slightly more often, with a bias of approximately $51/49$.)

TIP

To find out the real chance of a two-outcome situation, you can repeat the situation many, many times and see how often each outcome appears; this will estimate the value of *p* in the binomial distribution, and you can figure out probabilities for n = 1.

Applying the Wrong Probability Distribution

If you are working on a probability problem that involves a probability distribution, your best bet is to make sure you are starting on the right foot, going in the right direction, setting your compass to north, and all of that. Bottom line: Choose the right distribution, or you'll probably get in trouble right away and not be able to go anywhere — or you'll get somewhere, but it won't be the right direction, and you'll get the wrong answer. Let's figure out how to prevent both situations.

REMEMBER

Start every probability problem by labeling every piece of information you are given in the problem, and also label what you are trying to find. That will help give you clues as to what the distribution is.

Once you've labeled the pieces you are given in the problem, you'll be able to pick up clues as to what distribution you are working with. Look for words and numbers that identify the distribution. For example, if you are given a sample size n, like 20, and a probability of an outcome, like 0.10, and you are repeating an event over and over and counting the number of times that outcome occurs, you've got yourself a binomial distribution. If you have a fixed amount of time instead of a fixed number of repetitions, yet you are still counting the number of times the outcome occurs in that time period, it's a Poisson.

TIP

Making an "if-then" sheet can be a big help in determining the correct distribution. Label the first column "If." That's where the clues go, such as, "If you are given n and p and you are counting the number of yeses or successes . . . ." Then, in the second column, you say, "Then, it's a binomial." On the next line, you might write "If you have a fixed time period and are counting the number of events in that time period . . ." in the first column, and then in the second column, you'd write, "Then it's a Poisson." This can help you remember right away how to start a problem correctly, which goes a long way toward finishing it correctly. You can then add a third column that contains an example of each type of problem. This is also a much better way to study for an exam than endlessly paging through your notes.

Leaving Conditions Unchecked

Many distributions and techniques require conditions, and it's easy to overlook them. However, it's important to check conditions to make sure you've got the right technique going, that the technique is going to work right when you use it, and you avoid the loss of points on an exam. (If you're like me, every point counts.)

Some of the most common conditions are:

>> Checking the sample size when answering questions about the probability for $\overline{X}$, the sample mean, when you are starting with a non-normal distribution. In other words, when you are using the Central Limit Theorem. The larger the value of n, the better, but for most cases, n of at least 30 works well. If n is a low number like 6, you can't solve the problem using Z; you would have to use the t-distribution.

>> Checking $np \geq 10$ and $n(1-p) \geq 10$ before using the normal approximation to the binomial distribution. You have to check both conditions because if p is small, the first condition will be harder to meet since it's based on p, but the second condition will be easier to meet since it's based on 1 - p, which is large if p is small. And if p is large, 1 - p is small, so the second condition will be harder to meet. If p is ½, both conditions come up with the same values.

>> Checking to make sure you've got a normal distribution (or an approximate normal distribution) before using Z to solve the problem. The more distributions that come up in probability, the harder it is to distinguish them. But one of the most common ones is the normal, and you can't use Z (the standard normal) unless X starts out normal or approximately normal.

>> Checking to make sure the following four conditions are met for a binomial distribution:

- The sample (n) has a fixed number of trials or values.

- Yes or no are the only two outcomes.

- The trials are independent.

- The trials have the same probability of success or yes, which is p.

Each of the ten distributions discussed in this book has certain conditions to go with it. It would be a good idea to make a list of the distributions and their conditions and study them separately. Then, when you get to doing a problem, you won't forget to check the conditions.

Confusing Permutations and Combinations

Permutations and combinations are both counting rules, and they each have a special formula and a special use. But they can be confusing at times. Both occur when you are counting the number of ways to draw k objects out of n total objects without replacement (not putting them back after you take them out).

REMEMBER

The bottom line is that a *permutation* is the number of ways to choose k objects without replacement from n objects when the order matters. A *combination* is the number of ways to choose k objects without replacement from n objects when the order doesn't matter. The notation for a permutation is P_k^n, and the notation for a combination is C_k^n.

The formula for a permutation is $P_k^n = \dfrac{n!}{(n-k)!}$, where n! means taking the numbers down from n to n - 1 to n - 2 to . . . to 2 to 1, and multiplying them together. For example, 4! (pronounced "four factorial") is 4(3)(2)(1) = 24. 0! by convention is 1. Let's look at an example.

Find the number of ways to permute two objects taken from a total of four objects. To do this, you'd use $P_2^4 = \dfrac{4!}{(4-2)!} = \dfrac{4(3)(2)(1)}{2(1)} = 12$, which shows that there are 12 ways to do this. If your objects were labeled A, B, C, D, and you choose two of them and order matters, your 12 ways to do it would be AB, BA, AC, CA, AD, DA, BC, CB, BD, DB, CD, and DC.

You have n! in the numerator because you have n items and it's starting with the number of ways to rearrange those n items: n choices for the first object, n − 1 for the second object, n − 2 for the third object, and so on. But you are only rearranging k of the items, not n of the items, so you divide out the ways to rearrange the ones you left behind, because you don't want to count those. So in $P_2^4 = \frac{4!}{(4-2)!}$, you start 4! ways to rearrange all four items, then you take two out, and divide by $(4-2)!$, which is the number of ways to rearrange what you didn't take.

The formula for a combination is $C_k^n = \frac{n!}{k!(n-k)!}$. You start with the permutation formula, but you also divide by k!. Remember, order does not matter in a combination, so you also divide out the number of ways you rearrange the k items you did pick, since you don't care about their order. Let's look at an example.

Find the number of combinations of two objects taken from a total of four objects (where order does not matter). To do this, you'd use $C_2^4 = \frac{4!}{(4-2)!2!} = \frac{4(3)\,(2)\,(1)}{[2(1)]\,[2(1)]} = 6$, which shows that there are six ways to do this. If your objects were labeled A, B, C, D, and you choose two of them, and order doesn't matter, your six ways to do it would be AB, AC, AD, BC, BD, and CD.

If you calculate P_k^n and C_k^n using the same values of n and k both times, you'll have more permutations than combinations (unless k is 0 or 1, then it'll be the same). That's because order matters with a permutation and not with a combination.

The art is in choosing which one to use, the permutation or the combination. Ask yourself whether order matters. Here are a couple of examples:

>> How many ways can you arrange five people around a circular table?

>> How many ways can you form a committee of three people out of a team of six people?

To answer the first question, you ask whether order matters. Yes, order is the point of the problem because you are arranging the five people in different ways each time. Now you know it's a permutation, so you need to figure out n and k. The value of n is the total number of values, which is five, because you are working with all five, and all five are being chosen to be part of the arrangement, so k is also 5. In this case, you have $P_5^5 = \frac{5!}{(5-5)!} = \frac{5!}{0!} = \frac{5(4)\,(3)\,(2)\,(1)}{1} = 120$. That's a lot of ways to arrange five people!

To answer the second question, you have $n = 6$ and $k = 3$, and order doesn't matter because there are no jobs given on the committee; you're either chosen for the committee or you aren't. So you have $C_3^6 = \frac{6!}{(6-3)!(3!)} = \frac{6(5)\,(4)\,(3)\,(2)\,(1)}{3!\,(3)\,(2)\,(1)} = \frac{720}{(6)\,(3)\,(2)\,(1)} = 20$.

Assuming Independence

Independence of two events A and B means A and B don't affect each other. Think of two roommates, Bob and Moran. If Bob and Moran are truly independent, then whether Bob is in the apartment or not doesn't affect the chance that Moran is in the apartment. In other words, if B = Bob is in the apartment, and M = Moran is in the apartment, then $P(B|M) = P(B)$ and $P(M|B) = P(M)$.

Problems are easier to solve if you have independence, because, as you just saw, the conditional probability of $P(B|M)$ becomes a marginal probability, $P(B)$, which is easier to work with.

REMEMBER

The trouble is, folks like to assume by default that events are independent, and unfortunately, they are not independent unless it says so in the problem, or you can prove it using one of three ways: $P(A|B) = P(A)$; $P(A|B) = P(A|B')$; or $P(A \cap B) = P(A)P(B)$.

Here are a couple of scenarios to look at. Suppose $P(A|B) = 0.4$, $P(A) = 0.6$, and $P(B) = 0.2$. Are A and B independent? No. The first two probabilities given would have to be equal, and they are not. Next, suppose $P(A) = 0.5$, $P(B) = 0.4$, and $P(A \cap B) = 0.1$. Are A and B independent? No, because $P(A \cap B) \neq P(A)P(B)$.

Independence is a very important topic in probability, and always checking to see if events are independent can save you a lot of points on an exam. Events that are randomly chosen as a sample are independent, since random means that every sample of the same size (n) has the same chance of being chosen; their results wouldn't be able to affect each other.

TIP

There are independent events, where the intersection probability is the product of $P(A)$ and $P(B)$, as you just saw. There are dependent events, where the intersection probability is not the same as the product of $P(A)$ and $P(B)$. And there are, if you will, "ultra dependent" events, whose intersection probability is zero. These are mutually exclusive. If Bob and Moran are mutually exclusive roommates, it's a bad deal because one will leave the apartment if the other one is there.

Chapter **20**

Ten Probability Distributions to Compare

You've practiced with over ten different probability distributions in the chapters of this book: The discrete uniform, binomial, normal, normal approximation to the binomial, Poisson, geometric, negative binomial, hypergeometric, continuous uniform, and exponential. If you are working on a probability problem that involves a probability distribution, your best bet is to make sure you start with the correct distribution; otherwise, you'll likely head in directions you never intended and end up with the wrong answer. In this chapter, you get a refresher on each of the probability distributions covered in this book so that you can prevent both situations.

Discrete Uniform Distribution

If you have values of X that take on values of consecutive integers from a to b, inclusive, and each possible value of X has an equal probability, then you are dealing with the discrete uniform distribution. One easy example of such a distribution is the roll of a fair die; each outcome has probability ⅙.

Suppose that you decide to choose your ice cream cone flavor at random. Ten flavors are available, and of those ten, two are sugar-free (flavors 9 and 10). What's the chance you randomly select a sugar-free ice cream flavor?

In this case, the only two parameters are: a = the starting point for X, and b = the ending point for X. Here, a = 1 and b = 10. We know this problem is a discrete uniform distribution because of the integers being used, along with the fact that each number has the same chance of occurring. Here, each ice cream flavor has $\frac{1}{10}$ chance of being selected. To solve this problem, you want to find $P(X \geq 9)$ using the discrete uniform probability mass function (pmf). This function is $\frac{1}{(b-a+1)}$ for any value because each value has the same probability. For ten flavors of ice cream, $P(x) = 1 \mid (10-1+1) = \frac{1}{10}$. So $P(x \geq 9) = P(X = 9) + P(X = 10) = \frac{1}{10} + \frac{1}{10} = \frac{2}{10}$.

The mean of the discrete uniform is $\frac{(b+a)}{2}$, the midpoint and the variance is $\frac{(b+a+2)(b-a)}{12}$, and the standard deviation is the square root of that, which is $\sqrt{\frac{(b+a+2)(b-a)}{12}}$.

TIP

The discrete uniform distribution can take on negative values as well as positive values or zero. For example, imagine a six-sided die whose numbers are 1, 0, -1, -2, -3, -4. Each probability is still the same — one over the number of outcomes — which is ⅙ in this case.

Binomial Distribution

If you are given a sample size or number of trials along with a probability of a certain event that is deemed a "yes" or a "success," then you are likely dealing with a binomial distribution. The binomial distribution has two parameters indicated by letters: *n*, for the sample size, and *p*, for the probability of success (or yes, which means the characteristic or event you are interested in).

For example, Frederick tests boards to see whether they are defective or okay to use when building a house. He randomly selects 10 boards, and the usual rate of defectives is 5 percent. What is the probability that Frederick finds fewer than two defective boards?

In this problem, you see the two fixed values that describe the population, known as *parameters*. Those parameters are n, the sample size, and p, the probability of a "yes." In this example, n = 10, p = 0.05, and "yes" is the chance of getting a defective board. You are counting defective boards, so in this case, a success is not a good thing, but it doesn't have to be. A success is whatever you are counting; that's why the word "yes" is often used instead of "success" to denote it. Now you know you are using the binomial distribution with X = number of yeses, and you use the binomial table or the binomial probability mass function to answer Frederick's question: What is $P(X < 2)$?

To solve this problem, you can use the pmf and find $P(X = 0) + P(X = 1)$, or you can use the binomial table (Table A-1) in the appendix (see Chapter 8 for more on using the binomial table). Using the table you get $P(X < 2) = P(X \le 1) = 0.914$.

Normal Distribution

A normal distribution is signified by the bell-shaped curve. When a problem requires a normal distribution, you're usually given the mean and standard deviation to go with it. The *mean* tells you where the center of the bell-shaped curve is, and the *standard deviation* tells you the amount of variability that exists in the curve. The mean, notated by μ, and the standard deviation, notated by σ, are the two parameters of the normal distribution. So, there aren't any real clues to look for beyond that.

Suppose that X = the number of national exam scores that have a normal distribution with a mean of 75 and a standard deviation of 5. What's the chance that a randomly selected student scored more than 85? In this case, you have a normal distribution with $\mu = 75$ and $\sigma = 5$, and you want to find $P(X > 85)$.

To answer this question about the national exams, you want $P(X > 85)$. You change X to a Z value on the standard normal distribution using the formula, $Z = \dfrac{X - \mu}{\sigma}$, and you use the Z table to solve. (See Chapter 9 for more information.) For this example, you would find $P(Z > 85) = P\left(Z > \dfrac{85 - 75}{5}\right) = P(Z > 2) = 1 - 0.9772 = 0.0228$ (see Figure 20-1).

FIGURE 20-1:
Solving a
problem that
requires the
normal
distribution.

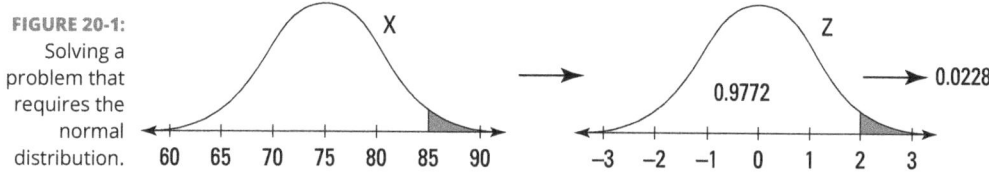

Normal Approximation to the Binomial Distribution

If you are starting with a binomial distribution with n and p, and n is large enough so that $np \ge 10$ and $n(1-p) \ge 10$, it's likely too large of a problem for the binomial distribution or the binomial probability mass function to solve. In these cases, what you do is use the normal distribution to approximate it.

For example, suppose that you are checking the graduation rates of college basketball players in Ohio. You pick 100 players at random. The reported rate of graduation is 80 percent. You want to know the chance that more than 85 of these players graduated.

This problem calls for a normal approximation to the binomial. To identify that this is what you need to do, you notice that you have n and p, something you are counting ("yes" = graduated), and that n is large (here it is 100, which at least seems large). Then you check to see if the two conditions are met (100 times $0.80 = 80$ is at least 10, and 100 times $0.20 = 20$ is at least 10), so yes, the conditions are met. Because they are met, you change X to a Z from the standard normal distribution by subtracting the mean of the binomial, np, and dividing by the standard deviation of the binomial, $\sqrt{np(1-p)}$ to get $Z = \dfrac{X - np}{\sqrt{np(1-p)}}$, and solving from there. (See Chapter 10 for details.)

WARNING

Don't try to use the regular binomial probability mass function to solve problems where the conditions for np and $n(1-p)$ are met (in other words, when n is so large it meets these two conditions). It will take so long for you to work out the problem that you won't have time to finish the other problems you are responsible for. And the normal approximation is very close when it comes to getting the exact answer. So it's a win–win.

For example, suppose that $n = 100$ people were surveyed in the United States, and we know that 70 percent of people in the United States have Internet access. You want to know the probability that your sample shows more than 73 of them have Internet access. This is a binomial situation with $n = 100$ and $p = 0.70$. You want to find $P(X > 73)$, where X = the number of people in the sample that have Internet access. If you do this problem using the binomial, first, $n = 100$ won't be on your binomial table because it's too large. Second, you'd have to find $P(X > 73) = P(X = 74) + P(X = 75) + P(X = 76) + \ldots + P(X = 99) + P(X = 100)$, each with the probability mass function for the binomial, which as you may recall is $\binom{n}{x} p^x (1-p)^{n-x}$. This would require a great deal of work to calculate out. So, the normal approximation is the better way to go here.

How to solve this problem using the normal approximation to the binomial? You start by checking the conditions. You have $n = 100$ and $p = 0.70$. You want $P(X > 73)$. The conditions to check are np and $n(1-p)$ are both at least 10. We have that $100(0.70) = 70$ and $100(0.30) = 30$, which are both at least 10, so the conditions are met for the normal approximation.

Next, you form a Z statistic using $Z = \dfrac{X - \mu}{\sigma}$, where $\mu = np$ from the binomial mean $(100)(0.70) = 70$, and $\sigma = \sqrt{np(1-p)}$ from the binomial standard deviation, $\sqrt{100 * 0.70(1-0.70)} = 4.58$. So $Z = \dfrac{X - 30}{4.58}$. You want $P(X > 73) = P\left(Z > \dfrac{73 - 70}{4.58}\right) = P\left(Z > \dfrac{3}{4.58}\right) = P(Z > 0.66) = 1 - 0.7454 = 0.2546$. This is an approximate answer.

Poisson Distribution

The Poisson distribution counts occurrences of an event over a fixed time period or over a fixed area of space. It sounds like a binomial in a way, but it's completely different. The time period or the space is fixed, you are counting how many occurrences (X) happen, and X is a whole number from 0 to infinity.

The conditions for X to be a Poisson are:

>> X counts the number of events or occurrences within a specified time or space.

>> The events occur independently of each other.

>> No two events can happen at the same time or place.

Some examples of a Poisson distribution include the number of chocolate chips in a 2-inch-wide chocolate chip cookie; the number of defective spots in a 10×10 square-foot carpet; and the number of accidents that occur at a certain intersection over a ten-day period.

The probability mass function of the Poisson distribution is $\dfrac{e^{-\lambda}\lambda^x}{x!}$, for x = 0, 1, 2, 3, and so on.

The average (or mean) rate of occurrence of the events over the fixed time or space is λ (pronounced "lambda"). That's the parameter for the Poisson. So if X is the number of cars that roll through a certain stop sign in an hour and X is Poisson with a mean of 6, $p(x) = \dfrac{e^{-6}6^x}{x!}$, where x = 0, 1, 2, 3, and so on. (Recall that e is the natural logarithm and goes into your calculator. We know e^1 is approximately 2.72.) If you want $P(X=8)$, then you put 8 in for X in the pmf and you get $\dfrac{e^{-8}6^8}{8!} = \dfrac{(0.000335)(1,679,616)}{40,320} = 0.013974$.

The mean of a Poisson distribution is λ. The variance of X when X is Poisson is also λ, which means that as the mean increases, so does the variance of the Poisson. The standard deviation is $\sqrt{\lambda}$.

If you change units of time or space, the value of the mean and standard deviation change with them. This is called a *Poisson process*. For example, if X = the number of occurrences of someone rolling through a red light in one hour, and the mean is $\lambda = 6$, then Y = 2X is the number of occurrences of someone rolling through a red light in two hours, and the mean is $2\lambda = 12$.

If λ is large enough that calculations get out of hand, you can approximate the Poisson with the normal distribution. The larger λ is, the better job it does, but $\lambda = 20$ is thought of as a good enough value to use the normal approximation. You change X to Z by subtracting the mean of the Poisson (λ) and dividing by the standard deviation of the Poisson, $\sqrt{\lambda}$, so you have $Z = \dfrac{X - \lambda}{\sqrt{\lambda}}$ and you solve from there as you would any Z problem.

Geometric Distribution

The geometric distribution enables you to model and find probabilities for X = the number of trials needed until the first "success," or "yes," occurs. For example, how many flips of a coin do you need to get the first head? If you get TTTTTTTTTH, then X = 10. Or, how many hours go by at an intersection before someone walks across? If it takes 3 hours before the first person walks across, then X = 3 because the first two hours were failures and the third hour was a success.

REMEMBER

The conditions a geometric distribution must meet are: (1) You have a sequence of independent trials of some random process; (2) each outcome is either a yes or a no (or success or failure); (3) the probability of success is the same for each trial and is denoted by p; and (4) X counts the number of trials *up to and including* the first success (which means the number of failures prior to the first success is X − 1).

The geometric distribution feels and looks a bit like a binomial distribution, except for the trials. They are not fixed. You don't know how many trials you'll need before the first success; in fact, that's what the random variable X stands for.

The formula for the pmf of the geometric distribution is $P(x) = p(1-p)^{x-1}$, where $x = 1, 2, 3$ and so on. This looks like the geometric series for those who have had calculus, raising a fraction — a number like $(1-p)$ — to higher and higher powers, hence the name "geometric distribution." You can see in the formula that the X − 1 failures, each one with probability $1-p$, come out to $(1-p)^{x-1}$. The first success has probability p, and you put that in front to get $P(x) = p(1-p)^{x-1}p$. This looks similar to a binomial distribution, but there is no combination in front to count the number of ways to get the successes and failures, because there is only one way in the geometric: The failures all come first, then the successes.

For example, suppose that you are conducting quality control for a lightbulb factory. Five percent of the bulbs you test are defective. What's the chance that the tenth bulb is the first defective one you find? You have p = 0.05, and you want $P(X = 10) = (1-0.05)^9(0.05) = 0.03$.

The mean and variance of the geometric are mean $= \dfrac{1}{p}$ and variance $= \dfrac{1-p}{p^2}$. The standard deviation is $\sqrt{\dfrac{1-p}{p^2}}$. The mean makes sense, because, say the chance of winning an instant lottery ticket is $p = \frac{1}{100} = 0.01$. How many tickets do you *expect* to go through in the long run until you win? $\frac{1}{p} = \frac{1}{0.01} = 100$. (Something to remember when purchasing lottery tickets.)

Negative Binomial Distribution

The negative binomial distribution is, in a way, the opposite of the binomial distribution, hence its name. Instead of fixing the number of trials and counting the number of successes, you fix the number of successes and count the number of trials up to and including that number of successes. You basically count how long it takes in a number of trials to get k successes, where *k* is determined ahead of time. For example, how many times do you have to flip a coin until you get five heads? Or how many cars do you have to see go through the intersection before you find two that roll through it without really stopping?

TIP

Some of the distributions may start to sound similar at this point; the differences are subtle when reading the problem, but huge when coming up with the right solution. It's always a good idea to have a list of all the distributions with their conditions, including what X is doing, and their probability mass functions (except in the case of a normal distribution, which is continuous, and you use a table because the probability density is too much).

The conditions for the negative binomial distribution are the following:

» You observe a sequence of independent trials from some random process.

» You classify the outcome of each trial into two groups: success or failure (or "yes" and "no").

» The probability of successes, p, is the same for each trial.

» X counts the total number of trials *up to and including* the kth success. (That means there will be $n - k$ failures.)

The pmf for the negative binomial is $P(x) = \binom{x-1}{k-1} p^k (1-p)^{x-k}$, where $x = k, k+1, k+2$, and so on, where $k > 0$. You are basically filling in the last spot with the last success (because then you're finished), and then finding the number of ways to rearrange the rest of the $k - 1$ successes. You have k successes, each with probability p (hence p^k), and $x - k$ failures, each with probability $(1-p)$, hence the $(1-p)^{x-k}$. The value of x starts at k because to get k successes, you need at minimum, k trials.

Let's say Praveen is calling people on a list to do a survey, and the chance that someone answers is 0.90. (Praveen will follow up with those who don't answer later.) What is the chance he makes five calls before he gets the third answer?

This scenario has a negative binomial distribution because you are counting the number of trials (phone calls), and you have $p = 0.90$ as the chance for success (someone answers). Note the way this problem is worded: He makes five calls and *then* he gets the third answer, so you want the probability that $X = 6$ calls up to and including the third answer ($k = 3$). So you want

$$P(6) = \binom{6-1}{3-1} 0.9^3 (1-0.9)^{6-3} = \binom{5}{2} 0.9^3 (0.1)^3 = \frac{5!}{2!(5-2)!} 0.000729 = 0.00729.$$

The mean of the negative binomial distribution is $\frac{k}{p}$, where k is the number of successes.

So in Praveen's case, the expected number of calls before the third person answers is $\frac{3}{0.9} = 3.33$. (That's why our probability of taking six calls was so small.) The variance of the negative binomial is $\frac{k(1-p)}{p^2}$.

Hypergeometric Distribution

The hypergeometric distribution is one in which you are using combinations such as C_m^n to find probabilities. For this distribution, you write the notation differently; you write C_m^n as $\binom{n}{m}$, where you have n items total, and you are choosing m of them without replacement and where order does not matter.

These are the five conditions for the hypergeometric distribution:

>> You sample without replacement from a population of N total. (The capital N stands for the population size, not the sample size.)

>> Every individual in the population has an equal chance of being sampled.

>> You classify the total population into two groups: Those with the characteristic of interested, called the *marked population,* and everyone else. The marked population could be those who own a pet, trees that need to be cut down, or people with a certain disease.

>> The total number of marked individuals is M, and the total population size is N; both M and N are fixed.

>> X counts the total number of marked individuals in the sample of size n.

Suppose that you are choosing people for a committee and the marked population is women and the unmarked population is men. There are $N = 25$ individuals, with $M = 10$ women, and $N - M = 15$ (the rest) are men. Suppose that you randomly select five people out of the entire group (so $n = 5$), and they are all women. X counts the number of women, so the value of X is 5. You want to know the probability that $X = 5$. You have a hypergeometric distribution on your hands.

TIP

The hypergeometric distribution is for situations in which you have one group, you are splitting the group into two groups in a certain way, and you want to know the probability of the outcome you'll get.

The pmf of the hypergeometric distribution is $P(x) = \dfrac{\binom{M}{x}\binom{N-M}{n-x}}{\binom{N}{n}}$. You have M individuals in the market population, and you are selecting x of them. You have the rest of the population that makes up $N - M$ individuals, and the rest of the sample comes from there ($n - x$ of them), and you're doing this out of N total individuals choosing n in the sample. The domain of x is $\max(0, M + n - N) \le x \le \min(M, n)$. (See Chapter 16 for more details.)

REMEMBER

The main difference between the hypergeometric distribution and other distributions is that you're reaching into the population all at once to get your sample; there are no individual trials.

The mean of the hypergeometric is $n * M/N$. The ratio of the marked population to total population is M/N, and you have a sample size of n.

The variance of the hypergeometric distribution is $\dfrac{nM}{N^2(N-1)}(N-M)(N-n)$. The standard deviation is the square root, which is $\sqrt{\dfrac{nM}{N^2(N-1)}(N-M)(N-n)}$.

Continuous Uniform Distribution

The continuous uniform distribution offers a break from some of the more complex distributions you have seen so far. Where the discrete uniform distribution can only take on integer values like -1, 0, 1, the continuous uniform distribution takes on those values plus all the real numbers between them (for example, $-1 \le x \le 1$). There are so many numbers between -1 and 1 when you include the real numbers that it's not possible to give every number a probability — you'd reach a total of 1 really quickly.

So, you can think of probability for the continuous uniform distribution as the area under the curve, $f(x)$, between two values, a and b, and as the probability that X is between a and b. This means X is a continuous random variable. And if that area is flat, or uniform, across the board for all values of X, you know X is a continuous uniform random variable on the interval from a to b.

For example, suppose that X is a continuous uniform distribution from 0 to 10. The value of $f(x)$ has to be $\frac{1}{10}$ so that the total area, the length $(b-a)$ times height, $f(x)$, equals 1. That's one of the rules of probability. If X is a continuous uniform distribution from 0 to 4, then $f(x) = \dfrac{1}{(b-a)} = \dfrac{1}{(4-0)} = \frac{1}{4}$ or 0.25.

TIP

The probability density function, $f(x)$, for the continuous uniform distribution is flat, straight across, so it makes a rectangle between a and b.

For example, suppose that you are waiting for a cab. Your wait time, based on previous days, has a continuous uniform distribution from 1 to 10 minutes. (After 10 minutes you will leave and cancel the cab.) What's the probability you wait more than 8 minutes?

Probabilities with continuous distributions typically involve the calculus topic of integration, which I don't go into here. However, when your density function looks like a rectangle, you use geometry to find the area under the curve that you want. In this case, you have $P(X>8) =$ length times height for the rectangle you need. Now the length is $10-8=2$ since X stops at 10 and starts at 8. And the height is the $f(x)$, which is $\dfrac{1}{(10-0)} = \frac{1}{10}$ or 0.10. So the area for $P(X>8) = 2*0.10 = 0.20$ (see Figure 20-2).

The mean of the uniform is $\dfrac{(a+b)}{2}$, which is the midpoint of the rectangle in terms of X. The variance is $\dfrac{(b-a)^2}{12}$.

FIGURE 20-2:
Finding
$P(X > 8)$ on a
continuous
uniform
distribution
from 0 to 10.

Exponential Distribution

The exponential distribution is a measurement usually involving time. It measures the time between Poisson occurrences, such as the time between sales on an Internet site, and it measures the length of time, such as the length of time a battery lasts, the time spent waiting in line at the grocery store, or the length of a helpline phone call.

The probability density function of the exponential distribution is $f(x) = \lambda e^{-\lambda x}$, where $x \geq 0$ (and zero otherwise). The parameter λ changes with the problem.

The mean of the exponential is $\frac{1}{\lambda}$, so if $\lambda = 5$, the mean of X must be $\frac{1}{5} = 0.20$. As λ gets larger, the expected value, $\frac{1}{\lambda}$, gets smaller, which is a characteristic of the exponential distribution.

REMEMBER (See Chapter 18 for more characteristics of the exponential distribution and their graphs.)

Because the exponential distribution is a continuous random variable, to find probabilities for it, you would typically integrate calculus; however, in this case, you have formulas for the three probabilities of interest: (1) $P(X < x)$; (2) $P(X > x)$, and (3) $P(a < X < b)$.

To find $P(X < x)$, you use the following formula: $P(X < x) = 1 - e^{-\lambda x}$. For example, Clint works at a helpline and X = the length of his calls in minutes. If the mean length of his calls is 20 minutes $= \frac{1}{\lambda}$, what's the probability that he talks less than 15 minutes during his next call?

Note that because the mean is 20 and is $\frac{1}{\lambda}$, that means the λ you use in the problem is $\frac{1}{20}$. You want $P(X < 15) = 1 - e^{-(1/20)(15)} = 1 - 0.47 = 0.53$. So there is a 53 percent chance that Clint will have a call that is less than 15 minutes.

Make sure the times are all in the same units throughout the entire problem. If the given mean is in minutes, and part of the problem talks about 1 hour, convert minutes to hours. If the given mean is in hours, and the problem talks about minutes, convert from hours to minutes. Remember the proportion $\frac{60 \text{ min}}{1 \text{ hr}} = \frac{30 \text{ min}}{x \text{ hr}}$ means $60x = 30$, and $x = \frac{1}{2}$ hour. Similarly, $\frac{60 \text{ min}}{1 \text{ hr}} = \frac{x \text{ min}}{1/2 \text{ hr}}$ means $60(\frac{1}{2}) = x$, so $x = 30$ minutes.

To find $P(X > x)$, you use the following formula: $P(X > x) = e^{-\lambda x}$. For example, Clint works at a helpline and X = the length of his calls in *hours*. If the mean length of his calls is 20 minutes, which in hours is ⅓, what's the probability that he talks more than 15 minutes during his next call? (Note that 15 minutes is equal to ¼ hour.)

Because the mean in hours is ⅓ and is $\frac{1}{\lambda}$, that means the λ you use in the problem is $\frac{1}{\frac{1}{3}} = 3$. You want $P\left(X > \frac{1}{4}\right) = e^{-(3)(1/4)} = 0.47$. So, there is a 47 percent chance that Clint will have a call that is 15 minutes (¼ hour) or less. Notice, this is the complement of the previous problem, which had him talking for less than 15 minutes.

TIP

Notice you can replace a less-than sign with a less-than-or-equal-to sign and get the same answer because $P(X = x) = 0$ in a continuous situation. Same with replacing a greater-than sign with a greater-than-or-equal-to sign. In the previous problem, we didn't worry about $P(X = 15 \text{ minutes})$ because that probability is zero in a continuous scenario like this one.

Appendix A

Tables for Reference

This appendix includes commonly used tables for finding probabilities for three important distributions: the binomial distribution, the normal distribution, and the Poisson distribution.

Binomial Table

Table A-1 shows the *cumulative distribution function* (cdf) for the binomial distribution (refer to Chapter 8). The cumulative probability is the total probability up to and including any given point. To use Table A-1, you need three pieces of information from the particular problem you're working on:

>> The sample size, n

>> The probability of success, p

>> The value of X for which you want the cumulative probability

After you have this information, find the portion of Table A-1 devoted to your n and look at the row for your x and the column for your p. Intersect that row and column, and you'll see the probability that X is less than or equal to your x. To get the probability of being strictly less than, greater than, greater than or equal to, or between two values of X, you manipulate the values of Table A-1, using the steps found in Chapter 8. For values of n less than 5, use the binomial formula to calculate your probabilities (see Chapter 8).

TABLE A-1   The cdf of the Binomial Distribution

Numbers in the table represent $P(X \leq x)$

n = 5

x	.01	.05	.10	.20	.25	.30	.40	.50	.60	.70	.75	.80	.90	.95	.99
								p							
0	.951	.774	.590	.328	.237	.168	.078	.031	.010	.002	.001	.000	.000	.000	.000
1	.999	.977	.919	.737	.633	.528	.337	.187	.087	.031	.016	.007	.000	.000	.000
2	1.000	.999	.991	.942	.896	.837	.683	.500	.317	.163	.104	.058	.009	.001	.000
3	1.000	1.000	1.000	.993	.984	.969	.913	.812	.663	.472	.367	.263	.081	.023	.001
4	1.000	1.000	1.000	1.000	.999	.998	.990	.969	.922	.832	.763	.672	.410	.226	.049
5	1.000	1.000	1.000	1.000	1.000	1.000	1.000	1.000	1.000	1.000	1.000	1.000	1.000	1.000	1.000

n = 6

x	.01	.05	.10	.20	.25	.30	.40	.50	.60	.70	.75	.80	.90	.95	.99
								p							
0	.941	.735	.531	.262	.178	.118	.047	.016	.004	.001	.000	.000	.000	.000	.000
1	.999	.967	.886	.655	.534	.420	.233	.109	.041	.011	.005	.002	.000	.000	.000
2	1.000	.998	.984	.901	.831	.744	.544	.344	.179	.070	.038	.017	.001	.000	.000
3	1.000	1.000	.999	.983	.962	.930	.821	.656	.456	.256	.169	.099	.016	.002	.000
4	1.000	1.000	1.000	.998	.995	.989	.959	.891	.767	.580	.466	.345	.114	.033	.001
5	1.000	1.000	1.000	1.000	1.000	.999	.996	.984	.953	.882	.822	.738	.469	.265	.059
6	1.000	1.000	1.000	1.000	1.000	1.000	1.000	1.000	1.000	1.000	1.000	1.000	1.000	1.000	1.000

n = 7

x	.01	.05	.10	.20	.25	.30	.40	.50	.60	.70	.75	.80	.90	.95	.99
								p							
0	.932	.698	.478	.210	.133	.082	.028	.008	.002	.000	.000	.000	.000	.000	.000
1	.998	.956	.850	.577	.445	.329	.159	.063	.019	.004	.001	.000	.000	.000	.000
2	1.000	.996	.974	.852	.756	.647	.420	.227	.096	.029	.013	.005	.000	.000	.000
3	1.000	1.000	.997	.967	.929	.874	.710	.500	.290	.126	.071	.033	.003	.000	.000
4	1.000	1.000	1.000	.995	.987	.971	.904	.773	.580	.353	.244	.148	.026	.004	.000
5	1.000	1.000	1.000	1.000	.999	.996	.981	.937	.841	.671	.555	.423	.150	.044	.002
6	1.000	1.000	1.000	1.000	1.000	1.000	.998	.992	.972	.918	.867	.790	.522	.302	.068
7	1.000	1.000	1.000	1.000	1.000	1.000	1.000	1.000	1.000	1.000	1.000	1.000	1.000	1.000	1.000

n = 8

x	.01	.05	.10	.20	.25	.30	.40	.50	.60	.70	.75	.80	.90	.95	.99
								p							
0	.923	.663	.430	.168	.100	.058	.017	.004	.001	.000	.000	.000	.000	.000	.000
1	.997	.943	.813	.503	.367	.255	.106	.035	.009	.001	.000	.000	.000	.000	.000
2	1.000	.994	.962	.797	.679	.552	.315	.145	.050	.011	.004	.001	.000	.000	.000
3	1.000	1.000	.995	.944	.886	.806	.594	.363	.174	.058	.027	.010	.000	.000	.000
4	1.000	1.000	1.000	.990	.973	.942	.826	.637	.406	.194	.114	.056	.005	.000	.000
5	1.000	1.000	1.000	.999	.996	.989	.950	.855	.685	.448	.321	.203	.038	.006	.000
6	1.000	1.000	1.000	1.000	1.000	.999	.991	.965	.894	.745	.633	.497	.187	.057	.003
7	1.000	1.000	1.000	1.000	1.000	1.000	.999	.996	.983	.942	.900	.832	.570	.337	.077
8	1.000	1.000	1.000	1.000	1.000	1.000	1.000	1.000	1.000	1.000	1.000	1.000	1.000	1.000	1.000

n = 9

x							p								
	.01	.05	.10	.20	.25	.30	.40	.50	.60	.70	.75	.80	.90	.95	.99
0	.914	.630	.387	.134	.075	.040	.010	.002	.000	.000	.000	.000	.000	.000	.000
1	.997	.929	.775	.436	.300	.196	.071	.020	.004	.000	.000	.000	.000	.000	.000
2	1.000	.992	.947	.738	.601	.463	.232	.090	.025	.004	.001	.000	.000	.000	.000
3	1.000	.999	.992	.914	.834	.730	.483	.254	.099	.025	.010	.003	.000	.000	.000
4	1.000	1.000	.999	.980	.951	.901	.733	.500	.267	.099	.049	.020	.001	.000	.000
5	1.000	1.000	1.000	.997	.990	.975	.901	.746	.517	.270	.166	.086	.008	.001	.000
6	1.000	1.000	1.000	1.000	.999	.996	.975	.910	.768	.537	.399	.262	.053	.008	.000
7	1.000	1.000	1.000	1.000	1.000	1.000	.996	.980	.929	.804	.700	.564	.225	.071	.003
8	1.000	1.000	1.000	1.000	1.000	1.000	1.000	.998	.990	.960	.925	.866	.613	.370	.086
9	1.000	1.000	1.000	1.000	1.000	1.000	1.000	1.000	1.000	1.000	1.000	1.000	1.000	1.000	1.000

n = 10

x							p								
	.01	.05	.10	.20	.25	.30	.40	.50	.60	.70	.75	.80	.90	.95	.99
0	.904	.599	.349	.107	.056	.028	.006	.001	.000	.000	.000	.000	.000	.000	.000
1	.996	.914	.736	.376	.244	.149	.046	.011	.002	.000	.000	.000	.000	.000	.000
2	1.000	.988	.930	.678	.526	.383	.167	.055	.012	.002	.000	.000	.000	.000	.000
3	1.000	.999	.987	.879	.776	.650	.382	.172	.055	.011	.004	.001	.000	.000	.000
4	1.000	1.000	.998	.967	.922	.850	.633	.377	.166	.047	.020	.006	.000	.000	.000
5	1.000	1.000	1.000	.994	.980	.953	.834	.623	.367	.150	.078	.033	.002	.000	.000
6	1.000	1.000	1.000	.999	.996	.989	.945	.828	.618	.350	.224	.121	.013	.001	.000
7	1.000	1.000	1.000	1.000	1.000	.998	.988	.945	.833	.617	.474	.322	.070	.012	.000
8	1.000	1.000	1.000	1.000	1.000	1.000	.998	.989	.954	.851	.756	.624	.264	.086	.004
9	1.000	1.000	1.000	1.000	1.000	1.000	1.000	.999	.994	.972	.944	.893	.651	.401	.096
10	1.000	1.000	1.000	1.000	1.000	1.000	1.000	1.000	1.000	1.000	1.000	1.000	1.000	1.000	1.000

n = 15

x							p								
	.01	.05	.10	.20	.25	.30	.40	.50	.60	.70	.75	.80	.90	.95	.99
0	.860	.463	.206	.035	.013	.005	.000	.000	.000	.000	.000	.000	.000	.000	.000
1	.990	.829	.549	.167	.080	.035	.005	.000	.000	.000	.000	.000	.000	.000	.000
2	1.000	.964	.816	.398	.236	.127	.027	.004	.000	.000	.000	.000	.000	.000	.000
3	1.000	.995	.944	.648	.461	.297	.091	.018	.002	.000	.000	.000	.000	.000	.000
4	1.000	.999	.987	.836	.686	.515	.217	.059	.009	.001	.000	.000	.000	.000	.000
5	1.000	1.000	.998	.939	.852	.722	.403	.151	.034	.004	.001	.000	.000	.000	.000
6	1.000	1.000	1.000	.982	.943	.869	.610	.304	.095	.015	.004	.001	.000	.000	.000
7	1.000	1.000	1.000	.996	.983	.950	.787	.500	.213	.050	.017	.004	.000	.000	.000
8	1.000	1.000	1.000	.999	.996	.985	.905	.696	.390	.131	.057	.018	.000	.000	.000
9	1.000	1.000	1.000	1.000	.999	.996	.966	.849	.597	.278	.148	.061	.002	.000	.000
10	1.000	1.000	1.000	1.000	1.000	.999	.991	.941	.783	.485	.314	.164	.013	.001	.000
11	1.000	1.000	1.000	1.000	1.000	1.000	.998	.982	.909	.703	.539	.352	.056	.005	.000
12	1.000	1.000	1.000	1.000	1.000	1.000	1.000	.996	.973	.873	.764	.602	.184	.036	.000
13	1.000	1.000	1.000	1.000	1.000	1.000	1.000	1.000	.995	.965	.920	.833	.451	.171	.010
14	1.000	1.000	1.000	1.000	1.000	1.000	1.000	1.000	1.000	.995	.987	.965	.794	.537	.140
15	1.000	1.000	1.000	1.000	1.000	1.000	1.000	1.000	1.000	1.000	1.000	1.000	1.000	1.000	1.000

(continued)

n = 20

x	.01	.05	.10	.20	.25	.30	.40	.50	.60	.70	.75	.80	.90	.95	.99
								p							
0	.818	.358	.122	.012	.003	.001	.000	.000	.000	.000	.000	.000	.000	.000	.000
1	.983	.736	.392	.069	.024	.008	.001	.000	.000	.000	.000	.000	.000	.000	.000
2	.999	.925	.677	.206	.091	.035	.004	.000	.000	.000	.000	.000	.000	.000	.000
3	1.000	.984	.867	.411	.225	.107	.016	.001	.000	.000	.000	.000	.000	.000	.000
4	1.000	.997	.957	.630	.415	.238	.051	.006	.000	.000	.000	.000	.000	.000	.000
5	1.000	1.000	.989	.804	.617	.416	.126	.021	.002	.000	.000	.000	.000	.000	.000
6	1.000	1.000	.998	.913	.786	.608	.250	.058	.006	.000	.000	.000	.000	.000	.000
7	1.000	1.000	1.000	.968	.898	.772	.416	.132	.021	.001	.000	.000	.000	.000	.000
8	1.000	1.000	1.000	.990	.959	.887	.596	.252	.057	.005	.001	.000	.000	.000	.000
9	1.000	1.000	1.000	.997	.986	.952	.755	.412	.128	.017	.004	.001	.000	.000	.000
10	1.000	1.000	1.000	.999	.996	.983	.872	.588	.245	.048	.014	.003	.000	.000	.000
11	1.000	1.000	1.000	1.000	.999	.995	.943	.748	.404	.113	.041	.010	.000	.000	.000
12	1.000	1.000	1.000	1.000	1.000	.999	.979	.868	.584	.228	.102	.032	.000	.000	.000
13	1.000	1.000	1.000	1.000	1.000	1.000	.994	.942	.750	.392	.214	.087	.002	.000	.000
14	1.000	1.000	1.000	1.000	1.000	1.000	.998	.979	.874	.584	.383	.196	.011	.000	.000
15	1.000	1.000	1.000	1.000	1.000	1.000	1.000	.994	.949	.762	.585	.370	.043	.003	.000
16	1.000	1.000	1.000	1.000	1.000	1.000	1.000	.999	.984	.893	.775	.589	.133	.016	.000
17	1.000	1.000	1.000	1.000	1.000	1.000	1.000	1.000	.996	.965	.909	.794	.323	.075	.001
18	1.000	1.000	1.000	1.000	1.000	1.000	1.000	1.000	.999	.992	.976	.931	.608	.264	.017
19	1.000	1.000	1.000	1.000	1.000	1.000	1.000	1.000	1.000	.999	.997	.988	.878	.642	.182
20	1.000	1.000	1.000	1.000	1.000	1.000	1.000	1.000	1.000	1.000	1.000	1.000	1.000	1.000	1.000

n = 25

x	.01	.05	.10	.20	.25	.30	.40	.50	.60	.70	.75	.80	.90	.95	.99
								p							
0	.778	.277	.072	.004	.001	.000	.000	.000	.000	.000	.000	.000	.000	.000	.000
1	.974	.642	.271	.027	.007	.002	.000	.000	.000	.000	.000	.000	.000	.000	.000
2	.998	.873	.537	.098	.032	.009	.000	.000	.000	.000	.000	.000	.000	.000	.000
3	1.000	.966	.764	.234	.096	.033	.002	.000	.000	.000	.000	.000	.000	.000	.000
4	1.000	.993	.902	.421	.214	.090	.009	.000	.000	.000	.000	.000	.000	.000	.000
5	1.000	.999	.967	.617	.378	.193	.029	.002	.000	.000	.000	.000	.000	.000	.000
6	1.000	1.000	.991	.780	.561	.341	.074	.007	.000	.000	.000	.000	.000	.000	.000
7	1.000	1.000	.998	.891	.727	.512	.154	.022	.001	.000	.000	.000	.000	.000	.000
8	1.000	1.000	1.000	.953	.851	.677	.274	.054	.004	.000	.000	.000	.000	.000	.000
9	1.000	1.000	1.000	.983	.929	.811	.425	.115	.013	.000	.000	.000	.000	.000	.000
10	1.000	1.000	1.000	.994	.970	.902	.586	.212	.034	.002	.000	.000	.000	.000	.000
11	1.000	1.000	1.000	.998	.989	.956	.732	.345	.078	.006	.001	.000	.000	.000	.000
12	1.000	1.000	1.000	1.000	.997	.983	.846	.500	.154	.017	.003	.000	.000	.000	.000
13	1.000	1.000	1.000	1.000	.999	.994	.922	.655	.268	.044	.011	.002	.000	.000	.000
14	1.000	1.000	1.000	1.000	1.000	.998	.966	.788	.414	.098	.030	.006	.000	.000	.000
15	1.000	1.000	1.000	1.000	1.000	1.000	.987	.885	.575	.189	.071	.017	.000	.000	.000
16	1.000	1.000	1.000	1.000	1.000	1.000	.996	.946	.726	.323	.149	.047	.000	.000	.000
17	1.000	1.000	1.000	1.000	1.000	1.000	.999	.978	.846	.488	.273	.109	.002	.000	.000
18	1.000	1.000	1.000	1.000	1.000	1.000	1.000	.993	.926	.659	.439	.220	.009	.000	.000
19	1.000	1.000	1.000	1.000	1.000	1.000	1.000	.998	.971	.807	.622	.383	.033	.001	.000
20	1.000	1.000	1.000	1.000	1.000	1.000	1.000	1.000	.991	.910	.786	.579	.098	.007	.000
21	1.000	1.000	1.000	1.000	1.000	1.000	1.000	1.000	.998	.967	.904	.766	.236	.034	.000
22	1.000	1.000	1.000	1.000	1.000	1.000	1.000	1.000	1.000	.991	.968	.902	.463	.127	.002
23	1.000	1.000	1.000	1.000	1.000	1.000	1.000	1.000	1.000	.998	.993	.973	.729	.358	.026
24	1.000	1.000	1.000	1.000	1.000	1.000	1.000	1.000	1.000	1.000	.999	.996	.928	.723	.222
25	1.000	1.000	1.000	1.000	1.000	1.000	1.000	1.000	1.000	1.000	1.000	1.000	1.000	1.000	1.000

Normal Table

Table A-2 shows the cdf for the normal distribution (refer to Chapter 9). To use Table A-2, you need three pieces of information from the problem you're working on:

>> The mean of X (the given normal distribution), which is μ

>> The standard deviation of X, which is σ

>> The value of X that you want the cumulative probability for

After you have this information, transform your value of X to a z value by taking your value of X, subtracting the mean, and dividing by the standard deviation, using the formula $Z = \dfrac{X - \mu}{\sigma}$ (refer to Chapter 9). Then look up this value of Z in Table A-2 by finding the row corresponding to the leading digit before the decimal point and the first digit after the decimal point of Z, and the column corresponding to the second digit after the decimal point of Z. The probability you find represents the probability that Z is less than or equal to that value of Z. For example, for Z = 1.23, go to the "1.2" row and the "0.03" column, and you'll find the probability that Z is less than or equal to 1.23 (which is 0.8907). To get the probability of being greater than Z or between two values of Z, manipulate the values of Table A-2, using the steps in Chapter 9.

The cdf of the Z Distribution (the Z Table)

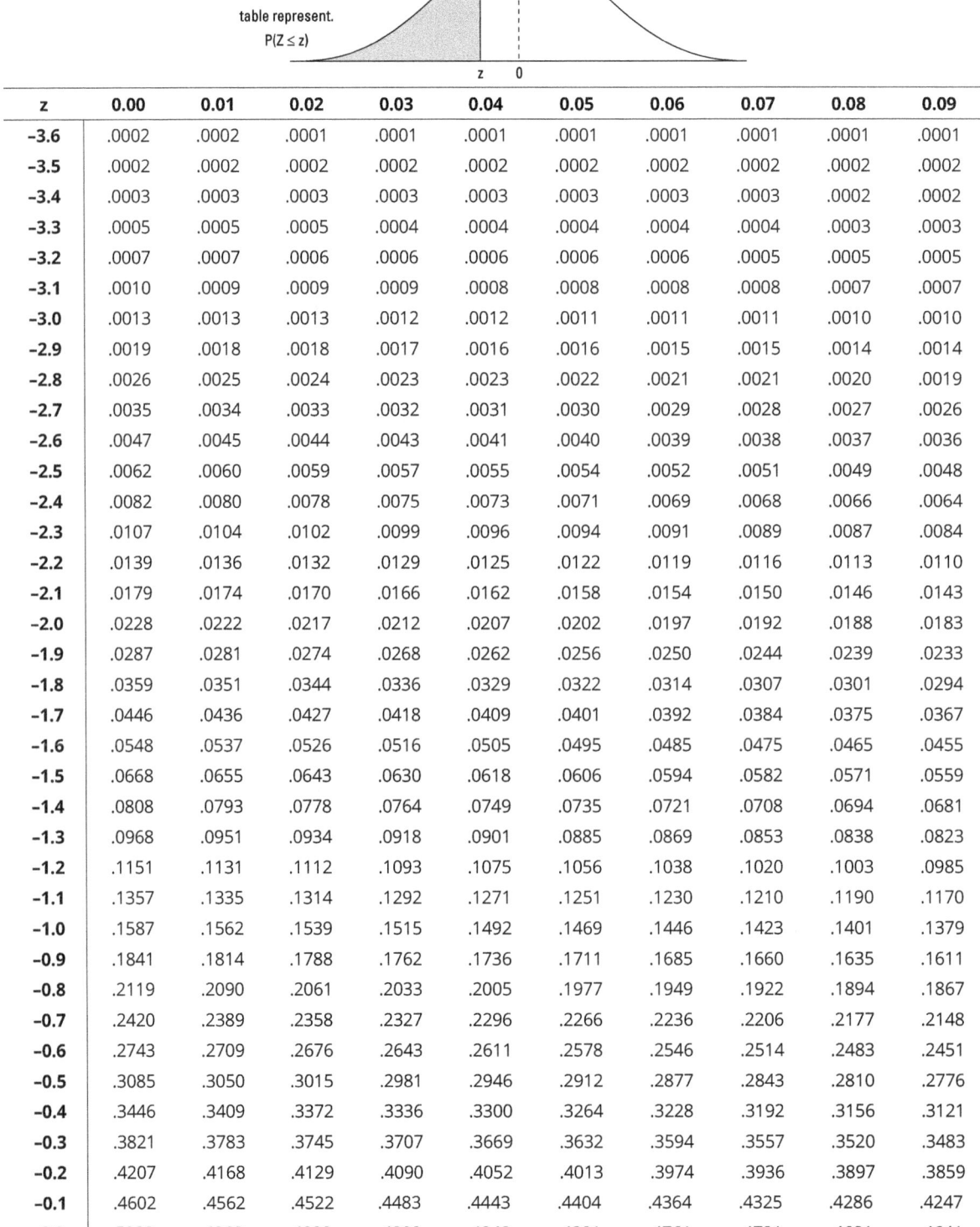

Numbers in the table represent.
$P(Z \le z)$

z	0.00	0.01	0.02	0.03	0.04	0.05	0.06	0.07	0.08	0.09
-3.6	.0002	.0002	.0001	.0001	.0001	.0001	.0001	.0001	.0001	.0001
-3.5	.0002	.0002	.0002	.0002	.0002	.0002	.0002	.0002	.0002	.0002
-3.4	.0003	.0003	.0003	.0003	.0003	.0003	.0003	.0003	.0002	.0002
-3.3	.0005	.0005	.0005	.0004	.0004	.0004	.0004	.0004	.0003	.0003
-3.2	.0007	.0007	.0006	.0006	.0006	.0006	.0006	.0005	.0005	.0005
-3.1	.0010	.0009	.0009	.0009	.0008	.0008	.0008	.0008	.0007	.0007
-3.0	.0013	.0013	.0013	.0012	.0012	.0011	.0011	.0011	.0010	.0010
-2.9	.0019	.0018	.0018	.0017	.0016	.0016	.0015	.0015	.0014	.0014
-2.8	.0026	.0025	.0024	.0023	.0023	.0022	.0021	.0021	.0020	.0019
-2.7	.0035	.0034	.0033	.0032	.0031	.0030	.0029	.0028	.0027	.0026
-2.6	.0047	.0045	.0044	.0043	.0041	.0040	.0039	.0038	.0037	.0036
-2.5	.0062	.0060	.0059	.0057	.0055	.0054	.0052	.0051	.0049	.0048
-2.4	.0082	.0080	.0078	.0075	.0073	.0071	.0069	.0068	.0066	.0064
-2.3	.0107	.0104	.0102	.0099	.0096	.0094	.0091	.0089	.0087	.0084
-2.2	.0139	.0136	.0132	.0129	.0125	.0122	.0119	.0116	.0113	.0110
-2.1	.0179	.0174	.0170	.0166	.0162	.0158	.0154	.0150	.0146	.0143
-2.0	.0228	.0222	.0217	.0212	.0207	.0202	.0197	.0192	.0188	.0183
-1.9	.0287	.0281	.0274	.0268	.0262	.0256	.0250	.0244	.0239	.0233
-1.8	.0359	.0351	.0344	.0336	.0329	.0322	.0314	.0307	.0301	.0294
-1.7	.0446	.0436	.0427	.0418	.0409	.0401	.0392	.0384	.0375	.0367
-1.6	.0548	.0537	.0526	.0516	.0505	.0495	.0485	.0475	.0465	.0455
-1.5	.0668	.0655	.0643	.0630	.0618	.0606	.0594	.0582	.0571	.0559
-1.4	.0808	.0793	.0778	.0764	.0749	.0735	.0721	.0708	.0694	.0681
-1.3	.0968	.0951	.0934	.0918	.0901	.0885	.0869	.0853	.0838	.0823
-1.2	.1151	.1131	.1112	.1093	.1075	.1056	.1038	.1020	.1003	.0985
-1.1	.1357	.1335	.1314	.1292	.1271	.1251	.1230	.1210	.1190	.1170
-1.0	.1587	.1562	.1539	.1515	.1492	.1469	.1446	.1423	.1401	.1379
-0.9	.1841	.1814	.1788	.1762	.1736	.1711	.1685	.1660	.1635	.1611
-0.8	.2119	.2090	.2061	.2033	.2005	.1977	.1949	.1922	.1894	.1867
-0.7	.2420	.2389	.2358	.2327	.2296	.2266	.2236	.2206	.2177	.2148
-0.6	.2743	.2709	.2676	.2643	.2611	.2578	.2546	.2514	.2483	.2451
-0.5	.3085	.3050	.3015	.2981	.2946	.2912	.2877	.2843	.2810	.2776
-0.4	.3446	.3409	.3372	.3336	.3300	.3264	.3228	.3192	.3156	.3121
-0.3	.3821	.3783	.3745	.3707	.3669	.3632	.3594	.3557	.3520	.3483
-0.2	.4207	.4168	.4129	.4090	.4052	.4013	.3974	.3936	.3897	.3859
-0.1	.4602	.4562	.4522	.4483	.4443	.4404	.4364	.4325	.4286	.4247
-0.0	.5000	.4960	.4920	.4880	.4840	.4801	.4761	.4721	.4681	.4641

Numbers in the table represent. $P(Z \leq z)$

z	0.00	0.01	0.02	0.03	0.04	0.05	0.06	0.07	0.08	0.09
0.0	.5000	.5040	.5080	.5120	.5160	.5199	.5239	.5279	.5319	.5359
0.1	.5398	.5438	.5478	.5517	.5557	.5596	.5636	.5675	.5714	.5753
0.2	.5793	.5832	.5871	.5910	.5948	.5987	.6026	.6064	.6103	.6141
0.3	.6179	.6217	.6255	.6293	.6331	.6368	.6406	.6443	.6480	.6517
0.4	.6554	.6591	.6628	.6664	.6700	.6736	.6772	.6808	.6844	.6879
0.5	.6915	.6950	.6985	.7019	.7054	.7088	.7123	.7157	.7190	.7224
0.6	.7257	.7291	.7324	.7357	.7389	.7422	.7454	.7486	.7517	.7549
0.7	.7580	.7611	.7642	.7673	.7704	.7734	.7764	.7794	.7823	.7852
0.8	.7881	.7910	.7939	.7967	.7995	.8023	.8051	.8078	.8106	.8133
0.9	.8159	.8186	.8212	.8238	.8264	.8289	.8315	.8340	.8365	.8389
1.0	.8413	.8438	.8461	.8485	.8508	.8531	.8554	.8577	.8599	.8621
1.1	.8643	.8665	.8686	.8708	.8729	.8749	.8770	.8790	.8810	.8830
1.2	.8849	.8869	.8888	.8907	.8925	.8944	.8962	.8980	.8997	.9015
1.3	.9032	.9049	.9066	.9082	.9099	.9115	.9131	.9147	.9162	.9177
1.4	.9192	.9207	.9222	.9236	.9251	.9265	.9279	.9292	.9306	.9319
1.5	.9332	.9345	.9357	.9370	.9382	.9394	.9406	.9418	.9429	.9441
1.6	.9452	.9463	.9474	.9484	.9495	.9505	.9515	.9525	.9535	.9545
1.7	.9554	.9564	.9573	.9582	.9591	.9599	.9608	.9616	.9625	.9633
1.8	.9641	.9649	.9656	.9664	.9671	.9678	.9686	.9693	.9699	.9706
1.9	.9713	.9719	.9726	.9732	.9738	.9744	.9750	.9756	.9761	.9767
2.0	.9772	.9778	.9783	.9788	.9793	.9798	.9803	.9808	.9812	.9817
2.1	.9821	.9826	.9830	.9834	.9838	.9842	.9846	.9850	.9854	.9857
2.2	.9861	.9864	.9868	.9871	.9875	.9878	.9881	.9884	.9887	.9890
2.3	.9893	.9896	.9898	.9901	.9904	.9906	.9909	.9911	.9913	.9916
2.4	.9918	.9920	.9922	.9925	.9927	.9929	.9931	.9932	.9934	.9936
2.5	.9938	.9940	.9941	.9943	.9945	.9946	.9948	.9949	.9951	.9952
2.6	.9953	.9955	.9956	.9957	.9959	.9960	.9961	.9962	.9963	.9964
2.7	.9965	.9966	.9967	.9968	.9969	.9970	.9971	.9972	.9973	.9974
2.8	.9974	.9975	.9976	.9977	.9977	.9978	.9979	.9979	.9980	.9981
2.9	.9981	.9982	.9982	.9983	.9984	.9984	.9985	.9985	.9986	.9986
3.0	.9987	.9987	.9987	.9988	.9988	.9989	.9989	.9989	.9990	.9990
3.1	.9990	.9991	.9991	.9991	.9992	.9992	.9992	.9992	.9993	.9993
3.2	.9993	.9993	.9994	.9994	.9994	.9994	.9994	.9995	.9995	.9995
3.3	.9995	.9995	.9995	.9996	.9996	.9996	.9996	.9996	.9996	.9997
3.4	.9997	.9997	.9997	.9997	.9997	.9997	.9997	.9997	.9997	.9998
3.5	.9998	.9998	.9998	.9998	.9998	.9998	.9998	.9998	.9998	.9998
3.6	.9998	.9998	.9999	.9999	.9999	.9999	.9999	.9999	.9999	.9999

Poisson Table

Table A-3 shows the cdf for the Poisson distribution (refer to Chapter 13). To use Table A-3, you need two pieces of information from the problem you're working on:

>> The mean of X (the given Poisson distribution), which is equal to λ

>> The value of X that you want the cumulative probability for

To use Table A-3, you find the column devoted to your value of λ and the row that represents your value of X. Intersect that row and column to find the probability that X is less than or equal to your value of X. To get the probability of being strictly less than X, greater than X, greater than or equal to X, or between two values of X, manipulate the values of Table A-3, using the steps in Chapter 13.

TABLE A-3   The Poisson cdf

Numbers in the table represent $P(X \leq x)$

		λ									
		.1	.2	.3	.4	.5	.6	.7	.8	.9	1.0
	0	.905	.819	.741	.670	.607	.549	.497	.449	.407	.368
	1	.995	.982	.963	.938	.910	.878	.844	.809	.772	.736
	2	1.000	.999	.996	.992	.986	.977	.966	.953	.937	.920
x	3		1.000	1.000	.999	.998	.997	.994	.991	.987	.981
	4				1.000	1.000	1.000	.999	.999	.998	.996
	5							1.000	1.000	1.000	.999
	6										1.000

		λ										
		2.0	3.0	4.0	5.0	6.0	7.0	8.0	9.0	10.0	15.0	20.0
	0	.135	.050	.018	.007	.002	.001	.000	.000	.000	.000	.000
	1	.406	.199	.092	.040	.017	.007	.003	.001	.000	.000	.000
	2	.677	.423	.238	.125	.062	.030	.014	.006	.003	.000	.000
	3	.857	.647	.433	.265	.151	.082	.042	.021	.010	.000	.000
	4	.947	.815	.629	.440	.285	.173	.100	.055	.029	.001	.000
	5	.983	.916	.785	.616	.446	.301	.191	.116	.067	.003	.000
	6	.995	.966	.889	.762	.606	.450	.313	.207	.130	.008	.000
	7	.999	.988	.949	.867	.744	.599	.453	.324	.220	.018	.001
	8	1.000	.996	.979	.932	.847	.729	.593	.456	.333	.037	.002
	9		.999	.992	.968	.916	.830	.717	.587	.458	.070	.005
	10		1.000	.997	.986	.957	.901	.816	.706	.583	.118	.011
	11			.999	.995	.980	.947	.888	.803	.697	.185	.021
	12			1.000	.998	.991	.973	.936	.876	.792	.268	.039
	13				.999	.996	.987	.966	.926	.864	.363	.066
	14				1.000	.999	.994	.983	.959	.917	.466	.105
	15					.999	.998	.992	.978	.951	.568	.157
	16					1.000	.999	.996	.989	.973	.664	.221
	17						1.000	.998	.995	.986	.749	.297
	18							.999	.998	.993	.819	.381
x	19							1.000	.999	.997	.875	.470
	20								1.000	.998	.917	.559
	21									.999	.947	.644
	22									1.000	.967	.721
	23										.981	.787
	24										.989	.843
	25										.994	.888
	26										.997	.922
	27										.998	.948
	28										.999	.966
	29										1.000	.978
	30											.987
	31											.992
	32											.995
	33											.997
	34											.999
	35											.999
	36											1.000

Index

Numerics

M

marginal probability, 47–49

mass function, probability

  geometric distribution, 174, 248

  hypergeometric distribution, 196–197

  Poisson distribution, 247

mean

  binomial distribution, 107–108

  continuous uniform distribution, 214–215

  exponential distribution, 227–228

  geometric distribution, 176–179

  negative binomial distribution, 188–190

  Poisson distribution, 164–165

  sample *see* sample mean

  of sampling distribution of $\bar{x}$, 137–138

misconception, 11–12

mistakes, ten probability, 235–242

  applying the wrong probability distribution, 239

  assuming independence, 241–242

  confusing permutations and combinations, 240–241

  forgetting where probabilities live, 235–236

  giving every two outcomes a 50–50 chance, 238

  leaving conditions unchecked, 239–240

  misinterpreting very small probabilities, 236

  thinking 1-2-3-4-5 and 6 can't win Powerball, 237

  thinking "I'm on a roll!," 237–238

  using probability for short-term predictions, 236–237

mutually exclusive events, 23–24

N

negative binomial distribution, 183, 248–249

  characteristics of, 184–185

  conditions for, 249

  mean, 188–190

  noting probabilities for, 186–188

  standard deviation of, 188–190

  variance, 188–190

normal approximation, 125–127

  to binomial distribution, 245–246

  Z statistic for, 128

normal distribution, 245

  basics of, 111–113

  bell-shaped pattern, 111

  cumulative distribution function (cdf) for, 259–261

  Poisson distribution with, 167–168

  probabilities for, 117–119

  solutions

    binomial distribution with, 132–133

    in normalizing, 122–123

  standard deviation, 111–113

normal table, 259–261

notation, 15

null set, 15

union probability, 16
  rule, 19

V

value expectation
  geometric distribution, 176–179
  hypergeometric distribution, 200–201
variable, continuous random, 207, 208
variance
  continuous uniform distribution, 214–215
  exponential distribution, 227–228
  geometric distribution, 176–179
  hypergeometric distribution, 200–201
  negative binomial distribution, 188–190
  Poisson distribution, 164–165
Venn diagrams, 29–32, 39–41

W

weather forecasting
  chance of rain, 8
  as long-term probabilities, 8
  meteorologist, simulation method, 10

Z

Z-distribution (Z-table), 114–116, 152, 259–261
Z-scores *see* Z-distribution (Z-table)
Z statistic, 128–129
Z-value, 152 *see also* Z-distribution

About the Author

Dr. Deborah Rumsey is a Fellow of the American Statistical Association and is a Professor of Teaching Practice at The Ohio State University in the Department of Statistics. She has taught over 40,000 students in her career, and loves teaching. Dr. Rumsey has also written several *For Dummies* books, including *Statistics For Dummies*, *Statistics II For Dummies*, and *Statistics Workbook For Dummies* (Wiley). She has won two teaching awards and is proud of the letters she gets from people who have succeeded in their classes after reading her books.

Dr. Rumsey lives on a ranch in Urbana, Ohio, with her husband. They have Aberdeen Angus cattle, goats, and one chicken. She enjoys traveling, spending time with her 23-year-old son, fishing, and cheering the Ohio State Buckeyes on to their next National Championship.

Dedication

This book is dedicated to my son, Clint. Every bet on you is a winner.

Author's Acknowledgments

I'd like to thank Elizabeth Stillwell for inviting me to write this book. Thanks to Katharine Dvorak, who was kind, patient, and helpful to me, and a great editor. I also want to thank the technical editor, Sara Conroy, for her good catches. Thanks to Kythrie Silva for being a faithful and supportive friend in any writing endeavor, and thanks to my husband, Eric, for cheering me on.

Publisher's Acknowledgments

Associate Acquisitions Editor: Elizabeth Stilwell
Senior Managing Editor: Kristie Pyles
Project Editor: Katharine Dvorak
Technical Editor: Sara Conroy
Editorial Assistant: Nina Hook

Production Editor: Magesh Elangovan
Cover Image: © Tran-Photography/stock.adobe.com